PRAISE FOR THE FLAG WAS STILL THERE

"E. LeBron Matthews has written a vivid, candid, humble account of his life as a combat Infantryman during the final stages of American involvement in the Vietnam War, 1970-1971. He describes the exhausting, demanding, sometimes mundane, but frequently terrifying, combat experiences as he and his "comrades in arms" fought an unpopular war and accomplished their missions, while helping each other stay alive in the process. His book is a great read—well written, thoroughly researched, very interesting, enlightening, and enjoyable."

—Major General William B. Steele,
U.S. Army (Retired)

"It's been fifty years since I was a twenty-year-old serving in the Army's 101st Airborne Division, 3/187th. Reading "The Flag Was Still There" triggered so many memories; the boredom punctuated by the occasional adrenaline rush, the sorrows, the tears and the forever brotherhood. Thank you LeBron for sharing this first-hand view of our struggles. A great testimony for future generations."

—Alan Davies, C/3-187 Infantry,
101 ABN, 1970-1971

"LeBron has put together a vividly clear and concise description of his military experience, one that every infantryman who served in Vietnam can relate to. We formed a brotherhood fifty years ago that only those who lived and worked together in the bush can truly understand. That bond continues to remain strong today. It was an honor to have served with LeBron and all of the other members of Charlie Company and the 3/187th Infantry."

— Russ Kagy, C/3-187 Infantry,
101 ABN, 1970-1971

"I've read and published many Vietnam books, and served there as a rifle platoon leader. LeBron has done an outstanding job describing the life of the Infantryman in his day to day life searching for the elusive enemy. You made me recall the day-to-day life of the Infantryman."

— Bob Babcock, B/1-22 Infantry,
4ID, 1966-1967

THE FLAG WAS STILL THERE

THE FLAG WAS STILL THERE

E. LEBRON MATTHEWS

Deeds Publishing | Athens

Copyright © 2020 — LeBron Matthews

ALL RIGHTS RESERVED—No part of this book may be reproduced in any form or by any electronic or mechanical means, including information storage and retrieval systems, without permission in writing from the authors, except by a reviewer who may quote brief passages in a review.

Published by Deeds Publishing in Athens, GA
www.deedspublishing.com

Printed in The United States of America

Cover design by Mark Babcock.

ISBN 978-1-950794-14-0

Books are available in quantity for promotional or premium use. For information, email info@deedspublishing.com.

First Edition, 2020

10 9 8 7 6 5 4 3 2 1

*To my wife, Pamela,
For our children, Anessia, Andrea, Audrey, and Andrew,
and all of our grandchildren*

*Dedicated to the memory of
Lt. Thomas W. Matthews, Jr.*

*And the other 58,177 men and women
who died in the Vietnam War.*

CONTENTS

Introduction	xi
Prologue	xix
1. Infantry Training	**1**
My Story	1
Number 98 in the Draft Lottery	5
Basic Training	16
Leadership School	28
Advanced Individual Training	31
Ten Days Leave	37
2. Red Warriors	**41**
Oakland, California (USA) to Long Binh (RVN)	41
Camp Radcliff and Fire Support Base Augusta	49
Qui Nhon	67
3. Rakkasans And Monsoon	**77**
Screaming Eagles	77
The Screaming Eagle Replacement Training Section	81
Rakkasans	83
Combat Operations in the Monsoon Season	86
January 1971	**115**
A New Year	115
The Climax of the Falls Operation	133

Dewey Canyon II/ Lam Son 719 (Part 1)	149
Dewey Canyon II	149
D-Day	163
Purple Heart Hill	175
The Star Spangled Banner	211
Dewey Canyon II/ Lam Son 719 (Part 2)	221
Sapper Attack	221
The Bunker Complex	242
Withdrawal from Khe Sanh	253
The Flats	265
Leave	265
Company RTO	276
Cobras Attack Down the Old French Highway	280
Typhoon Harriet	289
Division Color Guard	293
FSB Mexico AO	295
Happy Birthday, Zero!	295
Rubber Raft Patrols	296
Joyriding in a Loach	299
FSB Gladiator	301
The Mexico AO	302
R&R in Sydney	308
Miscellaneous	316
DEROS	318

Home	**321**
Fort Lewis and Home	321
Marriage	325
Back with the 4th Infantry Division	328
Granny's Death	332
Welcome Home	333
Photographs	**337**
Abbreviations, Acronyms, And Army Slang	**349**
Selected Bibliography	**363**
Military Records	363
Books	363
Periodicals	365
Internet Websites	366
Private Papers	367
About the Author	**369**

INTRODUCTION

Patriot Parade Field at Fort Benning, Georgia, is an extraordinary setting for soldiers. Its distinctiveness does not come from its appearance. It's just a grassy field with a reviewing stand on one side. Graduation ceremonies for basic training and infantry training classes routinely take place on the field with completion of every training cycle. Even so, the parade ground's uniqueness comes from what is not seen. Soil from battlefields of every war in which the United States Army has fought covers the ground on which these graduating soldiers march. Hence whenever today's soldiers step onto that field, they literally and metaphorically walk in the footsteps of American soldiers who fought on the battlefields of Yorktown, Antietam, Soissons, Pointe du Hoc, Soam-Ni, and LZ X-Ray.

LZ X-Ray was the first major battle of the United States Army in the Vietnam War. Over 2.7 million Americans served within the borders of South Vietnam during the war. Another 514,300 personnel were based in Thailand or on ships in the adjacent South China Sea. For the most part these men and women returned home to an ungrateful country. Their fellow citizens did not appreciate them or the job they did. These warriors did not march in victory parades. No public celebrations honored their service. Only decades later did people begin to express some gratitude for the service these venerable veterans rendered their country.

On June 5, 2010, the Army hosted a "Vietnam Veterans Welcome Home" ceremony on Patriot Parade Field. That Saturday morning over one thousand spectators watched a few hundred Vietnam veterans parade across the sacred dirt of Patriot Field. The aging veterans formed up in the National Infantry Museum's World War II Company Street. The Company Street adjoins Patriot Field opposite the reviewing stand. However, the ground between the field and the street gently slopes down six feet or more before leveling off between the old wooden World War II buildings. Therefore, the veteran's formation could not be seen by most of the spectators in the reviewing stand. When the veterans' formation crested the small rise and stepped onto the hallowed field, the crowd's thunderous applause drowned out the patriotic music of Fort Benning's army band. On both sides of the parade ground active duty soldiers stood at attention. Ragged ranks of old soldiers trooped between their two precise formations. For a moment, two living links in the chain that began on Lexington Green in 1775 overshadowed the symbolism of the soil. They represented two distinct generations of soldiers. One generation fought in the jungles of Vietnam; the other in the barren regions of Iraq and Afghanistan. Both were American soldiers. Both did their duty. Both deserve a grateful nation's respect.

The Vietnam veterans were no longer the vigorous warriors they had been in their youth. The years since the war had sapped their physical strength. Several cozied up to the formation in wheelchairs. Some crawled about in wheelchairs because of injuries received in Vietnam. Others simply had grown frail due to old age. Yet there was pride in these older men...pride about the service they rendered to their country in one of its most unpopular wars...pride in the belief they performed their duty and did their job well. But mostly it was pride for their comrades in the ranks. That pride constrained them to pay tribute to their buddies instead of tending to their own wellbeing. As the ceremony progressed, the hot southern sun beat down on their ranks. For a few, the summer heat proved too much. Intermittently an elderly body collapsed. Army medics quickly provided appropriate medical attention to the victim without interrupting the observance. These veterans had lived for decades without any recognition of their military service. Now finally the nation paid tribute to them. I was one of the veterans honored that day and the emotion I experienced in that celebration was inexpressible.

Some 2,709,918 Americans (approximately 9.7 percent of our generation) served in Vietnam.[1] Like the veterans of previous wars, our ranks grow thinner each day. The generations who fought at Yorktown, New Orleans, Chapultepec, Gettysburg, and San Juan Hill long ago became a faint memory. Recently the generation of our predecessors who struggled in the trenches of World War I disappeared. A few aged veterans from World War II and Korea remain with us, but not for long. In time, our generation too will be gone. Like earlier generations, our story deserves preservation for future generations.

In the summer of 2013, I went on vacation with my wife Pamela, daughter Audrey, and her daughter Eve. On the way home, we stopped by the Air Force Armament Museum at Elgin Air Force Base. An O-2A Skymaster forward air control airplane was on display. I shared an incident from my Vietnam experience involving a Skymaster. Audrey remarked, "I never knew that." Her comment made me realize that my family knew little about my Vietnam experience. A few weeks earlier, I had shared some stories with my grandson Peyton. He asked if he could record some of these experiences.

1.National Vietnam Veterans Foundation, Inc., www.nationalvietnamveteransfoundation.org/statistics.htm *[accessed August 20, 2015]*.

During the next couple of weeks, I pondered my family's interest in and ignorance of my army service.

1970 and 1971 were the most pivotal years of my life. In some ways, those two years confirmed and cemented core beliefs that I held long before Vietnam. In other ways, I came home a very different person. Many choices I made in life afterwards were rooted solidly in what happened during my tour of duty in Southeast Asia.

For many years, except to other combat veterans, I did not talk about Vietnam much. During that time, various triggers sparked flashbacks that led to strange behavior and despondent moods. As I grew older and our country's attitude towards Vietnam veterans changed, I began to converse about what happened. I discovered that sharing my past was therapeutic. Therefore, after pondering these things in the aftermath of Audrey's comment, I decided to put into writing for my family an account of my experience in Vietnam from September 14, 1970, to September 13, 1971. As I began writing, I started to assemble various documents, photographs, and memorabilia that I had cached in various locations. Studying these items released a flood of repressed memories and emotions as my mind wandered back to those enigmatic times. My attitudes and actions then were not always what today I wish they had been. Nonetheless, they are my story, for good or for bad.

I discovered that time really skews one's memory. Fortunately, several factors have helped me tell my story more accurately. First, I kept all of my Army records. These were especially useful in setting precise dates for certain events. I also have been able to consult other military reports and news articles. Second, I have numerous letters that I wrote to my parents from Vietnam. I also have an account of my tour that I wrote one week before my departure from Vietnam. Finally, my photographs from Vietnam not only preserved a visual record, but many had useful notations on the back. Even so, to some degree I still remained reliant on fallible reminiscences. Therefore certain events may be out of sequence. Some memories could not be checked for exact accuracy. Still they shaped my story. So I have included them in the narrative too. If there are errors or mistakes of facts, they are my responsibility alone.

I apologize to any of my comrades if my memory inexactly or wrongly marked their portrayal in this work. It was remembering your friendship from so long ago that was the greatest satisfaction from this effort. Eventually I completed a rough draft manuscript, or so I thought.

Then in July 2016, Russ Kagy from my platoon in Vietnam contacted me. He had recognized a photograph that I posted on the Internet. We exchanged photographs, emails, and talked on the telephone. He also consulted letters he wrote to his parents from Vietnam. Our communication clarified more of what had occurred in combat so long ago. I am very grateful to him and the bond we forged nearly five decades ago. We were able to pick up our friendship as if it were yesterday. Since then we have continued to communicate by telephone, email, and mail.

In January 2017, I also began communicating with CSM Dayton Herrington. Herrington was my company's First Sergeant in 1971. Even at age 86, his memory was sharp and he provided valuable insight into certain events in Vietnam. In April 2017, with the help of the West Point Association of Graduates, I made contact with Steve Metcalf who commanded my company March-July 1971. I was his RTO for much of that time. Since then I have also established communication with Alan "Uncle Al" Davies and Kevin Owens from my platoon. I am indebted to these brothers in arms for their invaluable memories, photographs, and perspectives, which provided me with a better overall picture of numerous events. In February 2018, CSM Herrington, Russ, Alan, and I attended the 101st Airborne Division

Association Snowbirds Reunion in Tampa, Florida. It was the first time that we all had been together since August 1971. Our face-to-face conversations provided even greater clarification of specific events as we identified our personal locations during those events. Combined together, our different perspectives completed the missing pieces of a jigsaw puzzle about what had happened so long ago. Since the reunion, the bond and communication between the members of first platoon of Charlie Company, 3rd Battalion, 187th Infantry Regiment, 101st Airborne Division (Airmobile) has only intensified.

As an ordained Southern Baptist minister, I have struggled with how to deal with the profanity that was integral to the Vietnam grunt's speech. These soldiers cussed profusely. The extent to which profanity was absorbed into my own speech shocked my family. One afternoon a few weeks after coming home, I was enjoying a delightful Sunday dinner at my parents' house. I turned to someone and said, "Pass the f***ing potatoes." My mother, a genteel Southern lady, almost passed out.

The Vietnam soldier's routine identification of certain equipment and realities often included frequent expletives. To omit their cursing would distort the reality of what a grunt's experience actually was like. Therefore, I have adopted the practice of printing the first letter of an expletive followed by asterisks for subsequent letters. Readers can then deal with the words according to their individual value system. Again, if the method offends anyone, I express some regret. Nevertheless, the method enabled me to tell the story correctly without compromising my current values. Profanity no longer is part of my daily vocabulary.

In the text of the narrative, I primarily have employed the acronyms and slang used by the troops in Vietnam. For the veterans of the war, these terms will be familiar. For those who did not serve, I hope that they will capture the flavor of the conflict. In some cases, I provided an explanation within the text. In other cases, the meaning of the acronym or slang is set forth in parenthesis after its first occurrence or in a footnote where the explanation might distract from the narrative. I also have included a glossary at the end of the book for the reader's convenience in recalling these terms.

I offer this book to my loving wife Pamela for her patience during those years when the war still haunted my memories so strongly. She was my reason to keep going while I was in Vietnam and has remained my motivation ever since. I am thankful for her and for the family that God granted us after I came home. I hope this narrative helps them to know me better.

This account is formally dedicated to Lt. Thomas W. Matthews Jr. and the other 58,177 men and women whose names are inscribed on the black stone wall of the Vietnam Memorial in Washington, D.C. They never came home from Nam to have families or to enjoy the families they already had. Therefore, it remains for us to give them their just legacy.

PROLOGUE

The man she fell in love with during the winter of 1969 and 1970 died in Vietnam. He was not killed in action, although he was a decorated combat infantryman. He came home alive with no physical scars. However, he was no longer the gentle, lighthearted boy who left her 12 months earlier. Something happened to him in Vietnam. Had the experience been too arduous? Perhaps, but thousands of other young men experienced similar — or worse — hardships in Southeast Asia. Still, something in him had changed in Vietnam. Had he seen too much brutality? Maybe, but thousands of other young men also saw the savage violence of ground combat in that war. Something in all of them surely died in the Vietnam War… something in him had died in Vietnam. He was both the man she loved and a complete stranger.

* * *

Our generation was known as baby boomers.[2] We grew up under the

---

2. In 1946, 3.4 million babies were born in the United States. That was 20 percent more than the previous year. In 1947 the number rose to 3.8 million. The boom continued until it reached 4 million by 1954. More than 4 million were born every

presidencies of Harry S. Truman, Dwight D. Eisenhower, and John F. Kennedy. By 1964, the 76.4 million "baby boomers" (born between 1946 and 1964) comprised almost 40 percent of the population of the United States.³

As children, we played outside with other children in our neighborhood, often staying out until dark or later. We watched Roy Rogers, Sky King, and the Mickey Mouse Club on black and white television sets. Most boys harbored a crush for Annette Funicello. This not so secret infatuation transmuted into adolescent lust when she exchanged her Mouseketeer ears for a bikini. We listened to rock and roll on transistor radios. We drove hot rods and muscle cars. The 1950s and early 1960s were a great time to be young.

The era also had its dark side, both domestically and internationally. Segregation separated "white" and "colored" in the American South. The African-Americans' struggle to attain equality often erupted in violence. Segregation politicians controlled most Southern state and local governments. They employed its power to suppress efforts to attain legal parity. Racist groups like the KKK overtly terrorized black communities. Nevertheless, inspired by heroic individuals such as Rosa Parks, guided by dynamic leaders such as Rev. Martin Luther King Jr., and backed by the full authority of the Federal Government, African-Americans slowly were gaining their long overdue civil rights.

The Cold War was punctuated with numerous terrifying days. The Soviet Union launched Sputnik, the world's first satellite, in 1957, and Americans panicked. In October 1962, the Cuban Missile Crisis pushed the world to the brink of nuclear Armageddon. Still for the

---

year until 1964 when the trend tapered off.

3. History. Baby Boomers, http://www.history.com/topics/baby-boomers [accessed November 18, 2014].

most part our lives remained isolated from anxiety. The greatest shock to our idyllic existence came when President Kennedy was assassinated in Dallas, Texas, on November 22, 1963. The following years brought escalation of the war in Vietnam. Nightly television newscasts brought the horror of the war into our homes. Our fairy tale lives shattered forever beneath the harsh pounding of these appalling realities.

My name is Edwin LeBron Matthews. I was O. H. "Jack" Matthews and Laura Frances Holland Matthews' first child. At the time of my birth, they lived with my paternal grandparents in Lakeview, Georgia. The closest obstetrical facility was in Baroness Erlanger Hospital, just across the state line in Chattanooga, Tennessee. I was born in that Tennessee hospital on a hot summer afternoon in July 1950. In my early formative years, my life revolved around family. Family life was not restricted to the nuclear family, but embraced the extended family of aunts, uncles, and cousins. In my childhood world cousins enjoyed a closer relationship than most siblings have today, especially within the sizeable Matthews clan.

My father was the youngest child of Myrtle Herbert Matthews and Serena Olive Anderson Matthews. Myrtle and Serena were married in 1907 and had seven sons and one daughter. Their thirty-three grandchildren called them "Pa" and "Ma." Pa was deceased when I went into the army.

My mother was the oldest child of Roy and Delia Holland of Chattanooga. Delia was a widow with one child, Harold Foster, when she married Roy. They had two other daughters. One died when she was an infant. The other, Katherine, was single and still lived with her parents. Their four grandchildren called Mr. and Mrs. Holland "Paw Paw" and "Granny."

The Matthews family was deeply rooted in the culture of the Cumberland Mountains in Tennessee and its heritage from the Old

South. Lunch was called "dinner" and what we now call dinner was known as "supper." I was taught to respect my elders and say "sir" or "ma'am." As a young boy, I absorbed the thrilling tales that Ma recounted of her family's life in the American Civil War. She could say "Yankee" in a malevolent tone that left no doubt where her family's loyalty had been in that disastrous war. Between her stories, the Civil War Centennial, and weekly visits to Chickamauga National Battlefield, I developed a keen interest in military history, especially the history of the Civil War.

Religion played an unquestioned role in the family's life. My parents held strong beliefs in a fundamentalist brand of the Christian faith. From birth, I grew up attending small conservative Southern Baptist churches. My paternal great-grandfather had been a preacher. My father, both grandfathers, and several uncles were Baptist deacons. Church attendance seemed as natural to me as school attendance.

I went by my middle name, which was to be an issue in a time when computers and paper forms were formatted only for the first name and middle initial. Consequently in the army—and later as a student at Georgia Tech—I frequently was known as Edwin or Ed rather than LeBron.

I had two younger sisters, Reita and Lisa. During my childhood, the family lived in Lakeview, Georgia; Knoxville, Tennessee; and Moncks Corner, South Carolina. When I was a sophomore in high school, we moved to Marietta, Georgia, where I graduated from Robert L. Osborne High School in 1968. Many of my classmates volunteered for service in Vietnam after graduation—the impact of the 1968 Tet offensive had not yet extinguished hope for an American victory in the Vietnam War. Others of us entered college and therefore were temporarily exempt from military service.

By the spring of 1970, no one talked of winning the war in Southeast Asia. Not the President. Not the Congress. Not the military. Not the press. Walter Cronkite even had said the war was unwinnable. The conversation had shifted from victory to withdrawal. *"How do we get out of Vietnam?"* Nevertheless, thousands of young boys still were being fed into the lethal crucible of the Vietnam War.

For the most part, the casualties were innocent young American teenagers. By this late in the war a majority of them had not volunteered. *Why die in a lost cause?* Most were forced into the military by the draft. Still, they answered their country's call to arms and served with honor. Being drafted rather than volunteering did not make their service any less worthy of their nation's gratitude. They were as courageous and worthy of honor as anyone who ever wears the uniform. They fought and died as American soldiers in a time when their country scorned their service. Yet their story almost universally has been omitted from history books and documentaries. One might think the war did not exist after 1969, but it did. American soldiers still fought a ground war in the rice paddies and jungles of Vietnam in 1970 and 1971. The story of these warriors—my story, for I was one of them—needs to be told and remembered. Monotony and boredom filled the daily existence of these grunts. The tedium, however, ended instantly when contact with the enemy was made. The war was real and still deadly. On average, over one hundred soldiers, sailors, and airmen died each and every week.

1. INFANTRY TRAINING

Drafted and Trained to be a Combat Infantryman

MY STORY

United States Army dogma maintains the infantry owns the last 100 yards of the battlefield. In his national bestseller about D-Day, Stephen Ambrose wrote, "The most extreme experience a human being can go through is being a combat infantryman."[4] While geographic location, uniforms, tactics, and technology change, the infantryman's experience remains universal. Combat infantry veterans of World War II, Korea, Vietnam, Panama, both Gulf wars, and Afghanistan share kinship in the fatigue, filth, and fright that are common to every conflict. All have been initiated into the exclusive fraternity of combat infantrymen. Its members hold a sense of superiority that other mortals cannot understand. It is not an arrogance born out of personal accomplishment. For inherent in membership is the belief that the best soldiers died on the battlefield. Consequently they grapple

---

4. Stephen E. Ambrose, *D-Day June 6, 1944: The Climatic Battle of World War II* (New York: Simon & Schuster, 1994), 419.

with the implication of this belief, a perplexity over why they, being an inferior soldier because they did not die, in fact lived. The combat infantry soldier saw war in its most primeval form and survived. Subsequently this peculiar pride in being an infantry soldier evaluates other combat arms as subordinate and habitually views all non-combat soldiers with outright disdain.

This is the story of my experience as a combat infantryman, as I remember it nearly five decades later. Some of my experience is shared by every infantry soldier since the campaigns of the ancient Egyptians. Some of it is characteristic of the Vietnam War. Some of it is unique to my story.

My story does not begin with my deployment to the Republic of Vietnam in 1970. Rather my Vietnam experience began with United States involvement in the Southeast Asian nation. I did not participate directly in these earlier events, but they shaped my perception of the incidents I did experience. Before the Tonkin Gulf episode, South Vietnam was an obscure name on a map barely studied in a Junior High School geography class.

The Eisenhower administration began sending United States military advisors to South Vietnam when I was in elementary school. These advisors started training Vietnamese ground combat units in 1960.[5] On August 2, 1964, North Vietnamese torpedo boats attacked the American naval destroyer U.S.S. *Maddox* in the Tonkin Gulf. A second attack was reported two days later. In response, President Lyndon B. Johnson ordered airstrikes against North Vietnam. In the aftermath of these events, Congress passed the Gulf of Tonkin Resolution and American military participation in the war escalated.

---

5. *The Encyclopedia of the Vietnam War: A Political, Social, & Military History*, 2000 ed., s.v. "Military Assistance and Advisory Group (MAAG), Vietnam" by Arthur T. Frame.

At the time, I paid little attention to the growing conflict in Vietnam. I was on summer break anticipating starting my freshman year at Berkley High School. School was scheduled to reopen in only a few days. In Moncks Corner, South Carolina, integration and the Civil Rights Act had more immediate impact on life than this distant war. For the first time in its history, a token handful of black students had enrolled in Berkley High. Already a host of federal, state, and local law enforcement officers rambled through the small community to insure integration went smoothly. During my sophomore year, my family relocated to Marietta, Georgia. Fitting in at a new high school troubled me more than the war in Southeast Asia. Berkley had been a football powerhouse. Osborne High School was a basketball school. Moncks Corner was a small rural town; Marietta was a mid-size urban city. As a shy adolescent boy, finding my niche in this new environment was extremely challenging. Although I was aware of the war in Vietnam, girls, cars, and sports occupied my thoughts.

Current events reports for civics and history classes in high school periodically forced me to examine specific aspects of the fighting in Vietnam. By my senior year, reports from Vietnam dominated every news media. The notorious Tet Offensive made ignoring the war impossible. For the first time, but not the last, names such as Hue, the Rockpile, and Khe Sanh entered my consciousness.

However, another factor made me increasingly aware of the war. Guys I knew personally were going to Vietnam. Several classmates from the class of 1968 at Osborne High School joined the Army or Marines immediately after graduation. In February, I heard one, Bill Milam, had been killed in action with the Marines.[6] Then Johnny

---

6. William Lawrence Milam III is listed as "Wilbur Lawrence Milam III" on the Vietnam virtual wall. http://www.virtualwall.org/dm/MilamWL01a.htm [accessed February 24, 2018]. Bill is listed as William in the *1968 Souvenirian*, vol. 23 of the

3

LaRue, a close friend at church, joined the Marine Corps and left for Vietnam. Vietnam no longer was just another daily report on the evening news. Vietnam had become a vast black hole, which was consuming my peers.

The summer after graduation, I turned 18 years of age. In compliance with Federal law, I registered with Selective Service. Kennesaw Junior College (now Kennesaw State University) had accepted my application to enroll as a college freshman for the fall quarter of 1968. Therefore, I received a 1-S student deferment. If I maintained satisfactory academic standards, I would be temporarily exempt from military service until I graduated college. However, at Kennesaw I was more interested in having a good time than studying. Therefore, because of poor academic performance during my first year of college, I lost that deferment for my sophomore year. The Vietnam War's potential to swallow me into that black hole was becoming far more likely. Consequently, I started taking closer note of what was occurring on the far side of the globe. In May 1969, the big war story in the news was the combat on Dong Ap Bia. For 11 days the 3rd Battalion of the 187th Infantry, 101st Airborne Division battled North Vietnamese Regulars for possession of the mountain. The struggle entered the lore of the famed 101st Airborne Division as the Battle of Hamburger Hill. Despite the announcement by the newly elected administration of Richard Nixon that American ground troops would be withdrawing from Vietnam, the war continued seemingly undiminished.

Osborne High School annual, 65. He was killed on February 10, 1969.

NUMBER 98 IN THE DRAFT LOTTERY

I was 19 years of age when I "won" the lottery. At the time, I still was enrolled in Kennesaw Junior College. However, I had ended my freshman year short of the required number of credits to keep a Selective Service 1-S student deferment. Therefore Selective Service reclassified me as 1-A. I took my Armed Forces Physical Examination on November 21, 1969. About that same time I received a letter from the Draft Board informing me that if I was "a full time satisfactory student" when I was ordered for induction, I would be eligible for a temporary 1-S(C) deferment until the end of the current academic year. At that time, I would need to have overcome previous deficiencies and have the required credits for two years of college.

Two factors led to me not pursuing an appeal with the Draft Board when I did receive my induction notice. First, I was uncertain that I could achieve the necessary academic standing for a successful appeal. Second, an appeal only seemed to defer the inevitable. Despite President Nixon's promise to end the war, its culmination was nowhere in sight in the spring of 1970. I developed a mindset of "Let's just get it over with as soon as possible so that I can get on with my life." Here I grossly miscalculated. The last draft call was held on December 7, 1972.[7] Had I succeeded in recouping my student deferment at the conclusion of the 1969-1970 academic year, I would have avoided military service entirely.

I was aware that at least two other options were available to escape compulsory military service. Both included severe consequences. First, a young man could attempt to evade the draft by going into

---

7. Selective Service website (http://www.sss.gov/lotter1.htm) [accessed June 2, 2017]. The authority to induct expired on June 30, 1973. Additional draft calls would have required an act of Congress.

hiding. This was a felony and if caught the draft evader would go to prison. The future of a convicted felon would be intensely restricted.

Second, one could flee to Canada. Thousands of draft dodgers did so. However, this risked becoming an expatriate for the rest of one's life. On January 21, 1977, President Jimmy Carter issued an unconditional pardon to the approximately 100,000 men who evaded the draft during the Vietnam War, allowing them to return to the United States without recrimination. This action was highly controversial. At the time public opinion strongly wanted to erase all memory of Vietnam. Many people unjustly impugned the men and women who had served there for the war itself. Even so, many Americans still opposed the pardon. Seven years previously, very few individuals could imagine that these draft dodgers ever would return to the United States.

Neither becoming a criminal nor abandoning my family and geographic roots appealed to me. Therefore, the military was the sole option left. If drafted, I dutifully would go into the armed forces of my country.

On December 1, 1969, Selective Service held the first draft lottery since World War II. The lottery determined the order for drafting men born between January 1, 1944, and December 31, 1950. Officials drew the 366 dates of a leap year from a large clear container. Each date was ascribed a number that corresponded to the order in which it was drawn. Men whose birthday fell on a given date would be called up for the draft according to the assigned number. Prior to this lottery, draft notices were issued based on "the oldest man first" method.[8]

The day after the lottery, Kennesaw Junior College posted its results on a bulletin board in the Student Center. The Student Center

---

8. Selective Service website (http://www.sss.gov/lotter1.htm) [accessed June 2, 2017].

was a large open room filled with folding tables encircled by stackable plastic chairs. Students regularly sat around the tables eating, studying, and chatting. The room contained more students than usual when I entered that fateful morning. Their voices added a chaotic roar to the morning's aberrant spectacle. A large bulletin board adorned with various student notices hung from the left-hand wall. Students rarely gave it more than a glance while passing. However, this Tuesday morning a mob of young men besieged the bulletin board. Every one of them searched the list of assigned numbers for their birthday. After discovering his number, each man stepped aside, allowing someone else to search. The expression on each face announced his fate. Big smiles telegraphed numbers unlikely to result in military service. Those slumping away with blank stares obviously had low numbers and academic problems.

I anxiously awaited an opportunity to check for my placement. Finally, I was close enough to read the notice. Starting at the top, I began scanning the list. The first date listed was September 14. That was not my birthday. I breathed a sigh of relief and continued to number two. April 22. I once more sighed. The pattern of fear followed by relief persisted as I continued perusing the list. When I saw "July 17," my heart seemed to stop beating. My assigned number was 98. There was little hope I would not be drafted. At the time no one knew exactly how many numbers would be inducted into the armed forces, but it almost certainly would be more than 98. In fact, by the end of 1970 numbers 1 through 195 would be called up for the draft.[9]

Because my number was so low, I expected to go into the military sometime that spring. At that time, I was unofficially engaged to Pamela Jean Coffman. Pamela was the only child of Hobert Carter

---

9. Selective Service website (http://www.sss.gov/LOTTER8.HTM) [accessed June 2, 2017].

"Coffy" Coffman and Charlotte Mayes Coffman. Mrs. Coffman had two children by a previous marriage, Barbara Richardson and Mike Strickland. Pamela attended Wills High School in Smyrna. Although she was only in the eleventh grade, we planned to get married after Pamela graduated. Rarely did the possibility that Vietnam might alter our plans enter into our blissful dreams. Yet occurrences such as my reclassification to 1-A and subsequently having to take a physical examination at the induction center occasionally signaled our wedding plans amounted to delusions. Nevertheless, for the most part, we managed to focus our deliberations on each upcoming weekend. Going to a movie or bowling were common pleasures on the Friday or Saturday nights when I was not working. After church on Sunday nights, we usually just ate at Old South Barbecue or the Dairy Queen in Smyrna. Overall, our dating life was simple and idyllic, more like that portrayed in a 1950s musical than in contemporary magazine articles about the wild lives of teenage hippies.

I worked part-time as a loading dock foreman for Rich's Department Store at Cobb Center on South Cobb Drive in Marietta. Cobb Center was the most prominent shopping mall in Cobb County. Rich's was an Atlanta icon in the twentieth century. Its downtown store opened just after the Civil War. By 1970, the retail chain had numerous branch stores in most of the major shopping malls around Metropolitan Atlanta.[10]

I was scheduled to work the afternoon and evening shift on Saturday, March 21. I frequently walked to work. The ambulatory commute took less than five minutes if I cut through the woods behind the mall. At the bottom of our driveway, I stopped by the mailbox.

---

10. In the late 1970s the local store chain merged with Federated Department Stores and eventually its identity was supplanted as its stores all converted to the Macy's brand.

Inside was an envelope addressed to me. The letterhead identified its source as the Selective Service. I did not bother opening the envelope. I already knew its contents. It was my draft notice. My only uncertainty was the exact date I would go into the military.

At the first opportunity that afternoon, I telephoned Pamela. Since I was working that night, we could not go out on a date. Nonetheless, I wanted to arrange for supper with her later that evening. I also needed to inform her about the Selective Service letter, but I was unsure how to tell her. She was vacuuming when the phone rang. When she answered it, I blurted out, "I'm going to Vietnam."

"No you're not," Pamela replied flippantly. She innocently assumed I was joking.

"Yes I am. I'm going to Vietnam," I coolly asserted. When she started to rebuff my bad sense of humor, I faintly whispered, "I got my draft notice today."

For a moment, silence intruded into our conversation. Then I could hear Pamela crying. Her tears continued as we finalized our dinner plans for that afternoon. After hanging up the telephone, I constantly glanced up at the clock above my desk on the loading dock. The agreed upon time for our dinner seemed reluctant to come. Finally the metal doors that separated the loading dock area from the sales floor swung open. I turned and saw Pamela striding straight towards me. I uneasily glanced back at the clock. She was a few minutes earlier than the designated time. I watched her approach, uncertain as what to say. Her eyes were red. She had been crying.

We held hands and walked to Dunaway Rexall Drugs across the mall from Rich's side entrance. Their soda fountain gleamed from polished chrome and shimmering plastic laminate. As usual on Saturday afternoons, the soda fountain was very crowded. Customers occupied every booth or table. We waited until two stools next to each other were available at the counter. After we ordered our meals,

I opened the letter. The notice was a standard form letter from the Selective Service. The first line read, "The President of the United States to." My name and address were inserted next. Pamela cried while I read, "Greetings: You are hereby ordered for induction into the Armed Forces of the United States, and to report…" When I finished, I handed her the letter. I suppose I did so to dispel any doubt about the future. The correspondence ordered me to report to the Armed Forces Examination and Entrance Station on Ponce de Leon Avenue in Atlanta at 6:30 a.m. on Wednesday, April 1, 1970. Dark lines of mascara traced the river of tears running down Pamela's cheeks. She could not control her emotion.

The waitress asked repeatedly, "Are you alright, honey?"

Pamela could not answer. She continued weeping hysterically. I briefly explained to the waitress that I just received my draft notice. The waitress grasped instantly that the notice probably meant I would be going to Vietnam. She said no more. Pamela laid her head on my shoulder and continued sobbing.

The Young People's Sunday School Department at Green Acres Baptist Church threw a going away party for me on Monday night before I left for the army. These young adults gave me a signed "Going to Miss You" card and a small New Testament. The New Testament was black leather with a metal plate in the front cover. They joked that if I carried it in my pocket, the plate might stop a bullet. I still have that New Testament. However, I ended up carrying a Gideon Bible instead. It was lighter and an infantry soldier quickly learns that every ounce matters. I did not carry my Bible in my pocket either. Vietnam's climate, especially during monsoon season, would have obliterated it.

* * *

Despite the reporting date, the letter was not an April Fools' joke.

Daddy and Mother drove me to the Induction Center early on the morning of April 1. Pamela snuggled tightly against me in the back seat of their tan Chevrolet Impala. It was still dark when the car pulled into the parking lot across the street from Atlanta's large brick Sears store on Ponce De Leon Avenue. The less impressive Armed Forces Examination and Entrance Station stood next to the store. An abnormally bleak gray sky condensed all sense of distance. A light rain fell. The asphalt and concrete caught the raindrops and formed shallow rivulets flowing downhill into the city gutters. Water puddled up on the pavement, coating streets with a glossy sheen. The moisture magnified the reflection of street lamps, traffic signals, vehicle taillights and headlights. So bright streaks of red, green, and white lights punctuated the darkness and created an enigmatic and dreary ambiance that was most appropriate for the occasion.

    I stayed in the car as long as possible. Our conversation was somewhat forced. The uncertainty of the future blanketed all of our thoughts. At the last possible minute, I kissed Mother and then Pamela good-bye. Neither was much of a kiss. I was too nervous and scared at that moment to permit any other emotion. Before I stepped out of the car, Pamela handed me a note to read later. She had written it the previous day in her third period class at school. "I LOVE YOU!" was emblazoned on the exterior fold in her handwriting. Inside the note apprised me that she had made the second highest grade in the class on her English test. The grade meant little to me, but I relished her expression of affection. I kept the note and read it repeatedly. I did not stop reading it until I received more current correspondence later.

    I slipped the note into my pocket and headed briskly across the street. A few streetlights and a traffic signal provided an eerie glow to the early morning. I stepped up onto the sidewalk. I turned left and started uphill to the old red brick building where I was to report. As I turned I glanced back in the direction of my parent's car. However, my anxiety blinded even a glimpse of my loved ones. I climbed the front steps of the Induction Center and opened the door. Inside I discovered a new and strange world.

* * *

Mother wrote a letter to her parents that afternoon:

> We are O.K. We took LeBron down this morning. He called about 12:20 and said he was going to Ft. Jackson, S. Carolina. This is at Columbia and won't be too far away. We sure are going to miss him but just hope he will be happy and satisfied.

When I discovered the letter just before Mother's death in 2013, I wondered if the phrase "hope he will be happy and satisfied" revealed her naiveté or was a veiled attempt to conceal reality.

* * *

Later that day all inductees formed in several straight rows. We then were instructed to take one step forward and raise our right hand. Thereafter I swore the following oath:

> *I do solemnly swear that I, Edwin LeBron Matthews, will support and defend the Constitution of the United States against all enemies, foreign and domestic; that I will bear true faith and allegiance to the same; and that I will obey the orders of the President of the United States and the officers appointed over me, according to regulations and the Uniform Code of Military Justice. So help me God.*

In the evening, a large number of draftees and a few volunteers boarded a Greyhound bus bound for United States Army basic training at Fort Jackson, South Carolina.[11] Around ten o'clock that night the bus reached the army post. The army fed the new arrivals mashed potatoes left over from the evening meal. They were cold and crusty, a harbinger of a disagreeable future.

One humorous incident occurred before that day ended. One of the inductees was a loud mouth, obnoxious braggart. All day he crowed about killing Viet Cong. The rest of us were too uneasy to talk much. Although the war certainly was on most of our minds, it was the last topic we wished to discuss. Mainly our conversations focused on our civilian lives in the recent past and the future we all hoped to enjoy one day. However, this guy's belligerent banter seemed to grow more bellicose and braggadocios as the hours wore on. By the time, we arrived at Jackson he had managed to ostracize himself from the entire group.

Upon arrival at Fort Jackson, the new recruits unloaded from the bus and went into an auditorium for an initial briefing. Recruits sat by assigned order in rows of metal folding chairs facing a stage at the front of the large room. Following the briefing, each row stood up sequentially and moved to the rear of the room. A medical team waited there to draw blood samples from each man. After giving blood, he went outside and walked around the building in order to reenter the room in the same order as we entered it initially. Inside again, each man sat down in his original chair. When the vociferous warrior got back to his seat, instead of sitting down, he passed out. His limp body hit the floor with a thud, sending metal chairs crashing into each other. The rest of us were too anxious to applaud, but an obvious

---

11. Department of the Army, Armed Forces Examining and Entrance Station, Atlanta, Georgia, Special Orders Number 65, 1 April 1970, Extract 2. TC 224.

snicker punctuated the culmination of banging chairs. I don't know what became of this man. I don't even know who he was. He in fact may have become a very good soldier or even a close friend. Nevertheless, in that moment it seemed to us that the prick of a tiny needle had dispensed a certain poetic justice to the self-proclaimed "hero."

* * *

At Fort Jackson's Reception Station, the men from Atlanta merged with recruits from other parts of the country. During our first four days in the army, army cadre harangued us as "receptees." Receptees had not yet metamorphosed into soldiers. According to army lore, receptees subsisted as the lowest life form on the planet. The army issued uniforms to us, gave us rudimentary instruction concerning military protocol, and tested everyone for aptitude to various military skills and assignments. The cadre continued to abuse us verbally and assign the most menial tasks to various receptees.

Early on the morning of our second full day in the army, a sergeant asked who brought a driver's license with them. He professed he needed drivers for a work detail. I wisely left my license at home when I reported to the induction center. However, many overconfident teenage boys were hesitant to part with this cherished badge of adolescence evolution. Accordingly, when the sergeant posed the question, they proudly raised their hands. The sergeant promptly formed them up and marched them away. To their mortification, they ended up "driving" wheelbarrows in the hot sun that day. However, escaping all work details was impossible. Everyone picked up cigarette butts and other trash several times each day.

The morning after arriving at Fort Jackson a sergeant instructed all new soldiers in basic drill movement. He covered rudimentary commands, various positions, and marching in formation. Then he marched us to a one story white wooden building for a haircut. The World War II era structure was in dire need of a fresh coat of paint. Moreover, "haircut" was a misnomer. "Sheep shearing" would be more descriptive. The barber shaved our heads with dull clippers that sometimes pulled out patches of hair rather than clip the locks. The surviving hair on top front of my head was about one and a half inched long. The rest was no more than a quarter of an inch in length. The barren spots eventually would grow back, but cultivating a new crop of hair takes prolonged time. Pamela cried when she first saw me a few weeks later.

Henceforth everyone would dress the same, in the exceedingly unfashionable olive drab fatigues, malformed baseball cap, and black leather combat boots. Consequently carrying the requisite paper work, receptees filed up to one end of a long counter inside an old warehouse. Each individual was handed an army green canvas duffle bag. As he proceeded down the counter, various articles of clothing were handed to him. The appropriate forms were checked upon receiving these items and he moved to the next station for the next piece of the uniform. The haircut and uniform were major components in erasing our civilian identity and assimilating us into "the big green machine" known as the United States Army.

During the time at the reception station, new soldiers received extensive aptitude testing to determine potential assignments. I scored very high on some of these tests and so I had to take some additional tests. One of these required going to a site some distance from the Reception Center. Consequently that evening the only transportation available to my quarters was walking. I was strolling alone down the sidewalk when the bugle announced the retirement of the colors. Being new to the army, my familiarity with bugle calls consisted of what I had heard in western movies or on television. I certainly did not know the army etiquette for that moment. Hence, I continued walking. Suddenly a shiny black car with a red flag on its hood pulled over and stopped. A single white star emblazoned the flag. A brigadier general got out of the vehicle and confronted me. Seeing the single star on his collar, I immediately snapped to attention and saluted (I already had learned that much). His expression and tone of voice broadcast his displeasure in my decision to continue walking. Upon learning that I was a new recruit, the general explained that when the national flag was lowered at the end of the day, soldiers were supposed to stop, face the flag, and salute until the bugle sounded its last note. I do not know who the general was, but I appreciated his attitude. He did not degrade me for my ignorance. He was the first superior not to do so since I arrived at Fort Jackson. Instead, he patiently instructed me in military decorum. Thereafter I never repeated that mistake.

We soon were promoted to "trainees" after receiving our initial army pay. In the army hierarchy, we remained at the bottom of the food chain, but we had crawled out from under our primordial rock and commenced our evolution into soldiers. The army assigned us to a basic training unit in order to undertake the next step in that evolution.

BASIC TRAINING

We traveled to our basic training unit in the back of a M35 army truck known as "the deuce-and-a-half." This two and one half ton cargo truck was the standard military vehicle for transporting equipment and personnel at the time. The truck bed had a wooden bench on each side. A canvas cover enclosed the bed, preventing its passengers from seeing anything during the trip.

When the driver finally pulled back the canvas back flap on the truck bed, its passengers rudely met the loudest and nastiest beast on the planet, the United States Army Drill Sergeant. These creatures brandished spit-polished combat boots, heavily starched fatigues, and a distinctive World War I era campaign hat. They were in our face screaming about what seemed to be the most innocuous infractions of military decorum. Other training cadre wearing shiny black helmet liners added their voices to the pandemonium.

Army regulations required fatigue pants legs to be tucked into the top of one's boots. One of my pants legs partially had come loose while climbing out of the truck and now hung down over the boot. A drill sergeant dispensed the most terrifying and profane reprimand I have ever received in my life. I can still see the veins in his neck bulging and his eyes popping out as he screamed, "You better write home and tell your momma to sell the s***house because you're a** is grass and I'm the lawnmower!" He then allowed me to work off any pent up rage this dressing-down had produced by quickly doing 25 pushups. These were the first of countless pushups I would do over the next eight weeks.

I was assigned to Company A, 3rd Battalion, 1st Training Brigade. Staff Sergeant Waters was the company's drill instructor. Waters was a light-skinned African-American. His uniform always had a starched crease in the trousers and was devoid of any hint of wrinkles. He wore his campaign hat cocked over the front of his face so as to obscure his eyes from our vision. Yet he saw every move his charges made. A thin burnt umber mustache rested securely over his upper lip. He was only about five foot six inches tall, but trainees were more afraid of him than anything else in the world. The punishment he meted out was harsh but effective. Pushups remained his most common chastisement. However, he became more creative with some transgressions. If he caught you with your hands in your pocket — an infraction of the military code of conduct — he put pinecones in that pocket. Their sharp spurs scrubbed the skin raw, instilling a habit of keeping hands out of the pocket. Other infractions drew comparable sentences.

His training discipline was equally tough. One night in late April or early May, the company finished a grueling night road march around one o'clock in the morning. Exhausted troops stowed their gear, took a quick shower, and hit the sack for a short night's sleep. We had been in bed about an hour when Waters came into the barracks blowing his whistle and banging his fist on the metal lockers. "Fall out for PT! Uniform of the day: army issue boxer shorts and t-shirts. You don't have those, you don't have nothing!" he screamed. Five minutes later, we were in the company street in our underwear doing PT (physical training) in the cool night air.

More than a few soldiers had replaced their army issue boxer shorts with civilian jockey shorts. At this point in our training program, any possession other than those items issued by the army was considered contraband and therefore subject to confiscation. So the forbidden shorts were forsaken and their owners fell into formation wearing nothing but combat boots (No, I was not one of them). We did pushups for the next hour. Regulations only permitted a maximum of 25 pushups in a single set. So Waters yelled. "Attention! Front lean and rest position! Move!" We then dropped into a prone position and knocked out 25 pushups. Then he screamed, "Attention!" We stood up and he gave the command, "At ease." A few seconds later he screamed, "Attention! Front lean and rest position! Move!" We executed 25 more pushups before we stood back up. Then we repeated the process time after time. I do not know how many push-ups we did that night, but my skinny arms were trembling when we stopped. Waters was exacting, but he turned us into soldiers and prepared us for the hard realities of war. I am alive today, in part, because of him.

I had scored high on all my aptitude tests. As a result, I was told that I could do anything I wanted in the army. Therefore, I selected what I believed were the safest jobs on the list. Naturally, the army in due course made me an infantryman instead.

Since my test scores were so high, I also was considered for an appointment to the United States Military Academy at West Point.[12] However, I needed at least one year of prep school prior to going to West Point. After graduation from the Academy, I would be required to serve one year of active duty for each year of education. That requirement equated to at least a ten-year obligation to the army in my case. I was a teenage boy in love and I longed to marry Pamela Coffman. Marriage would have to wait until after graduation if I went to West Point. I did not wish to wait five years to marry Pamela. I was certain waiting was unacceptable to her as well. Consequently, I soon declined the offer. Instead, I decided to attend OCS (Officer Candidate School) at Fort Benning, Georgia. This was the first of many decisions I made because of my unwavering longing to be with Pamela. In time, this yearning defined the essence of my will to live. In the despondent occasions I would face in the future, it drove my survival instinct and maintained a reason to stay alive. Too much credit for my survival cannot be ascribed to her. Because I loved Pamela Jean Coffman, I lived through the experience of Vietnam.

The Vietnam War was fought before the age of cell phones, the internet, and social media. Occasionally I called Pamela or my parents on a pay telephone near my barracks. However writing letters was our primary means of communication. Even prior to my military service, Pamela frequently wrote notes to me, especially while she was at school.

Feb. 10, 1970-3rd Period

Dear LeBron,
Hi, what cha doing? Did you study hard for the test you had

---

12. Disposition Form DA 2496, Reference AJJPC-RSP, 7 April 1970.

today. Was it easy? I sure am sleepy! I'm to [sic.] lazy to do any work so I decided to write ya a note. I'm in study hall.

These high school notes were usually on notebook paper. They often were embellished with hearts she had drawn and shorthand (which I couldn't read). Most were a single page, but occasionally she might write as many as 12 pages. A typical closing was

By the way what are we going to do this weekend? Gotta go for now!

Love ya always,
Mrs. E. L. Matthews to be

From the first week of basic training, Pamela and I established the habit of writing virtually every day. Postage on a letter at the time was only six cents for first class mail. Most of her letters were posted with a single stamp, but a few had six one-cent stamps. During the first few weeks I was away, Pamela's letters were on notebook or typing paper. Then she began using stationary. During basic training, she used pink. In AIT (Advanced Individual Training) she changed to pale blue and then yellow. Her letters usually began "Dear LeBron" or "Hi Honey!" More often, she used both. Some letters reflect the melancholy mood of the era. As many as 200 or more were killed and wounded in Vietnam each week.[13]

---

13. In 1969, 11,780 Americans died in Vietnam [http://www.archives.gov/research/military/vietnam-war/casualty-statistics.html (National Archives, Military Records website accessed June 2, 2017)]. In total 58,178 Americans died in the war [http://thewall-usa.com/summary.asp (The Vietnam Veterans Memorial website accessed May 30, 2019)].

> *This letter may not be too cheerful cos [sic.] I'm a bit upset at the moment. I was watching the Johnny Cash show and he was singing a song about Viet-Nam. Well, it got to me, but good. I tried not to think but things kept popping into my mind. I didn't want to listen but something was telling me that we would never get married. I really could not bare [sic.] the thought of losing you ... if anything ever happened to you, everyone might as well forget me too.*

In another letter, she wrote:

> *I saw a letter from you on the table by the door. I started reading it but had to leave the room before Mom came in because I didn't want her to see me crying. I was crying because I am happy. Ya know your [sic.] right. If we do put our trust in God everything will turn out alright. But I love you so much that I can't help worrying about you and what you are doing.*

Other topics were typical of any young couple in love in any age or circumstance. One described a chance encounter Pamela had with a girl I had once dated.

She is beginning to look bad. Her hair is growing out where she had put a lightener on it. Yuck!! She just didn't seem to look as pretty as she once did.

In that same letter she told me that another girl I dated had gotten married. Pamela's delight in both developments was obvious.

The letters habitually referred to our future marriage and family. They frequently mentioned our unfulfilled desire for physical intimacy. They all professed her love for me. Sometimes she expressed her sentiments in silly lingo.

"Guess what? Me WUV's u muches & muches."

At other times, it was more serious and direct.

I love this soldier guy. He's really cool. I love him with all my heart and soul. God has made our love possible... I love you.

Her letters inevitably ended "Love always, Pam" or "Love ya forever, Pam." Sometimes she might substitute "Mrs. Edwin LeBron Matthews" (or some variant of the name) for "Pam."

Although none of my correspondence with Pamela exists, it was equally intimate.[14] In one letter, she wrote the following:

> *The letters you have been writing me lately have really made me feel great. You don't know how much I like to hear you say, "Pam I love you." Just hearing(READING) that, I feel GREAT!*

Only one of Pamela's letters to me in Vietnam is existent. Likewise, all of my letters to her have been destroyed. Nevertheless, I cannot stress too much how significant this correspondence was to my mental health. It helped sustain my emotional and spiritual wellbeing during circumstances very alien to anything I previously had known.

* * *

My basic training battalion was located on the sloping hillside of Tank Hill, named for the 90 feet high silver cylindrical water tank on

---

14. A letter to my parents during basic training was written on stationary likely purchased at the PX (Post Exchange). The United States Army eagle is in the top left hand corner. A black line runs across the top of the page from behind the eagle to "UNITED STATES ARMY" in the top right corner. At the bottom left corner of the sheet is a figure of an infantryman advancing at port arms. Contemporary letters to Pamela almost certainly were written on the same stationary.

its crest.[15] The battalion buildings were 700 Series wooden structures erected at the beginning of World War II. These temporary structures were intended only to last until the end of the war in the 1940s. The barracks were all two story edifices designed to house 63 enlisted men and their noncommissioned officers. As war demand increased, bunk beds replaced the single beds, doubling the capacity.[16] Noncommissioned officers rarely lived in the barracks by the Vietnam era. Each floor was divided into a large open room and a latrine. Bunk beds lined both sides of the main room. Each bed was perpendicular to the outside wall. The row of beds alternated between the head or the foot of the bed being against the wall. Two army green wooden footlockers were placed by the end of bed in the central walkway of the room. This walkway was defined by the wooden columns that formed the structure of the building. Two metal lockers stood upright against the wall between beds. The spacing between beds was approximately three feet. Due to risk of fires each floor was required to have a "fire guard" posted during the night. The company orderly room, supply room, and mess hall were one-story structures adjacent to the barracks.[17]

---

15. Chris Rasmussen, April 20, 2011. "Post loses Tank Hill landmark," WWW.ARMY.MIL: The Official Homepage of the United States Army (http://www.army.mil/article/55258/post-loses-tank-hill-landmark/). The water tank was installed in March 1941 at a cost of $400,000. It held 1.8 million gallons and was 60-feet in diameter.

16. Amy Newcomb, February 2, 2012. "Look how far we've come: New Soldier barracks offers latest in comfort, privacy," WWW.ARMY.MIL: The Official Homepage of the United States Army (http://www.army.mil/article/73024/Look how far we've come New Soldier barracks offers latest in comfort privacy/).

17. The orderly room was the company office building. It contained offices for the company commander, first sergeant, and the clerk. The mess hall consisted of the

Life in a room with 50-60 male soldiers, most of who were in their late teens, didn't lack its frivolity. One young man had a tendency to roam the barracks at night. His wandering was not nefarious; he just kept people awake. One night while he was visiting a friend on the opposite end of the bay from his bunk, a couple of roommates short-sheeted his bed. An army bunk was covered with two white sheets and an olive drab wool blanket. For this prank, the top sheet was folded in half so as to make the bottom half of the top sheet appear to be the bottom sheet. Since the sheets of an army bunk were folded firmly under the mattress and covered tightly with the blanket, the fact that a person's feet could only extend down half the length of the bed was well disguised. That night the drill sergeant made a surprise inspection of the barracks. The roving soldier was a small, wiry creature. He raced across the room and slid into his bed before Sergeant Waters could flip on the lights. The distinctive sound of ripping cloth was heard in the darkness. No one laughed aloud until Waters departed. Then laughter resounded throughout the room. Though he never said so, the wandering victim of the prank likely had the cost of the torn sheet deducted from his meager wages.[18]

At roll call every morning soldiers were assigned work details. An entire platoon would be given the task of policing the ground outside the barracks. This consisted of lining up in a single row at one end of the company area and walking in line to the opposite side of the area. Along the way, each man picked up all trash and cigarette butts in his assigned lane. He also pulled up any grass or weeds that might be growing there — according to army logic the yard around the barracks should be dirt. Individual soldiers were given specific assignments,

---

kitchen and dining facilities. Equipment assigned to the company was stored in and issued out of the supply room.

18. An Army Private E-1 was paid $124.50 per month at that time.

such as cleaning the latrine or sweeping the barracks. Because I was considering West Point or OCS, I frequently missed such tasks in order to attend special meetings or training sessions. Unfortunately, I was not able to avoid KP (Kitchen Police). KP consisted of duty in the mess hall. Soldiers performed various tasks while on KP. Dishes and eating utensils had to be washed. Cooking pots and pans were washed by hand in a large sink in back of the kitchen. This was one of the most unpleasant tasks I did while in basic. Mopping floors, peeling potatoes, and serving meals also were part of the routine. Soldiers on KP were awakened long before the rest of the company and did not return to the barracks until late that evening. To soldiers the only congenial aspect of KP was it exempted one from the day's training. Ultimately, however, these missed skills still had to be learned or unfamiliarity with them might harm yourself or your buddies.

Basic training taught civilians how to be soldiers. Drills and the manual of arms taught teamwork and developed military bearing.[19] The physical nature of the curriculum honed the body's capabilities and prepared it for the strenuous demands of combat. What seemed at the time to be harassment, tuned soldiers' mental faculties and instilled the military discipline necessary on the battlefield.

During basic training, we were introduced to the M-16A1 assault rifle. The M-16 was the standard rifle of all army infantry units serving in Vietnam in 1970-1971. Special Forces tested the weapon in Vietnam during 1962. Units deploying to South Vietnam during the early years of the war carried the M-14. By 1966, these M-14s had been replaced by the M-16. The M-16 only weighed 6.5 pounds compared to the 8.7 pounds of the M-14. Its smaller size and the use of a plastic stock rather than wood accounted for the lighter weight. These

---

19. PAM 21-13, *The Soldier's BCT Handbook* (Washington, D.C.: Headquarters, Department of the Army, May 1969), p. 7.

design characteristics led to the derogatory nickname "the Mattel toy rifle."[20] The early M-16s were plagued by frequent jamming. However, a design change that added the forward assist assembly to push the bolt and a change in the gunpowder used in its ammunition largely corrected the problem. Despite these modifications the weapon's initial bad reputation never fully died out. Nevertheless, during the next 18 months this weapon would earn a special affection in my life. It never once failed to fire when I needed it to do so.

Our first day on the rifle range, the instructor began with a brief speech on the importance of learning how to use the weapon. He opened with a promise that we all were going to Vietnam. The only reason anyone might escape Vietnam was a war in Europe against the Soviet Union. So we would need to be really, really good to survive combat against the Russian horde. Then he callously announced four people had been killed by the Ohio National Guard at Kent State University and suggested we might have to fight in the United States.[21] Finally, he reminded us that Israel was always on the verge of a war and so we might be sent to the Middle East someday.

---

20. Mattel manufactured numerous toy guns during the 1950s and 1960s. These were some of the most popular toys among the generation that fought in the Vietnam War. So virtually every grunt had played with at least one Mattel gun in his childhood.

21. On May 1, 1970, demonstrations protesting the incursion into Cambodia commenced at Kent State University. The following day the ROTC building on campus was burned after a massive rally. On May 3, National Guard troops arrived on campus. The evening students gathering on the Commons were dispersed with tear gas. The next day the National Guard fired live rounds into a crowd of approximately 2,000 people. The firing only lasted about 13 seconds, but four people were killed.

THE FLAG WAS STILL THERE

* * *

While I was in basic training my first cousin, Tommy Matthews, was killed in Cambodia.[22] Tommy's death haunted my parents throughout my training and tour of duty in South Vietnam. No doubt, his death made the uncertainty caused by irregular mail and nightly television news even more traumatic for them. According to my sister, they were incessantly nervous the whole year that I was away in Vietnam.[23]

* * *

During basic training I developed a close friendship with John Little. John managed to get his car to Fort Jackson and toward the end of basic training, we received a weekend pass. John also was from the Atlanta area. Therefore, we decided to go home together. Technically the distance of the trip exceeded the geographical limits of our pass. However, we used some creative math to justify the excursion as being acceptable. Still if we had been caught, our logic undoubtedly would have failed to keep us out of trouble. John's car had an eight-track player but he only owned one or two cassettes for it. During the ride between Columbia and Atlanta, he listened to Johnny Cash sing "Ring of Fire" over and over and over… For the rest of our time at Fort Jackson, whenever we got a weekend pass, we headed home, listening to "Ring of Fire." There were other songs on the tape, but "Ring of Fire" was most memorable. I can still hear Johnny Cash's

---

22. 1st Lt. Thomas W. Matthews, Jr., Charlie Company, 2d Battalion, 12th Cavalry, 1st Cavalry Division. An account of his death is given by Keith William Nolan in *Into Cambodia: Spring Campaign, Summer Offensive, 1970* (Novato, California: Presidio Press, 1990), p. 350. Tommy was my father's brother Winfred's only son.
23. Reita M. King, e-mail message to the author, October 8, 2013.

deep, bass-baritone voice singing, "I fell into a burning ring of fire. I went down, down, down, and the flames went higher. And it burns, burns, burns, the ring of fire."[24]

LEADERSHIP SCHOOL

Basic training lasted eight weeks. After completing basic training, I was assigned to the Fort Jackson Leadership School for a two-week leader preparation course. Thereafter I was designated MOS 11Bravo and assigned to Company B, 11th Battalion, 3rd Brigade for AIT. MOS stood for Military Occupational Specialty. 11B10 designated me as a Light Weapons Specialist. In other words, the army needed me in the infantry. So ended the army's nefarious assurance given at the reception station that I could do any job I desired in the army. I might be able to do any job in the army, but that did not mean the army would allow me to do any job in the army. Infantry definitely was not one of the jobs I listed in my paperwork. Oh well, the man did say "I could do any job in the army." But he did not say "I would do any job I selected."

An infantryman in the Vietnam War was commonly called "a grunt." The origin of the slang identification remains unclear. The term was used in the early twentieth century for a low-level worker or laborer. Hence, the term "grunt work," referred to a job that was thankless, boring, and exhausting, but necessary. Although it is indisputable that infantrymen often engage in such work, there is no record of how this colloquial usage came to be connected to the infantry.[25] Another theory derives the terminology from vernacular for

---

24. Written by Merle Kilgore, June Carter. Lyrics © Shapiro Berstein & Co., Inc.
25. http://www.ask.com/government-politics/military-grunt-2f0faa7ac3711ceb.

a person with little or no authority in some organization. The word was applied to an infantryman due to his willingness to accept every situation and/or obey orders regardless of their consequences.[26]

The word's actual meaning denotes to a low, inarticulate, gruff and guttural sound that is made when a person lifts a heavy load. Perhaps then, the best theory is it was derived from lifting the heavy rucksacks infantrymen carried on their backs.[27] Whatever its origin, the infantry is a close-knit community and the slang word became a term of endearment that expresses mutual respect and shared experiences.

Soldiers at the Fort Jackson Leadership School enjoyed upgraded living conditions. The barracks was a new brick building as opposed to the ancient wood structures in which they lived during basic training. Instead of one large open bay, students were assigned to four man rooms on the second floor. The first floor contained classrooms, offices, and a day room. During the two weeks of school, my room earned a reputation for acquiring demerits. All four of us maintained very high training scores. However, we offset these with an excessive number of demerits. Unfortunately, many demerits were dispensed to the room as a whole rather than to an individual within the room. Some of these demerits were inexcusable. For example, one day the man responsible for emptying the trash can, simply turned it upside down with the trash still inside. All of us were cited for the infraction. Other demerits were caused by juvenile behavior. For instance, one night after lights out, a couple of my roommates instigated a pillow fight with an adjacent room. The fracas escalated and soon the entire barracks joined into the melee.

Consequently, the faculty grew particularly aggressive with the

---

26. http://onlineslangdictionary.com/meaning-definition-of/grunt.
27. Shelby L. Stanton, *Vietnam Order of Battle* (Washington, D.C.: U.S. News Books, 1981), p. 356.

four of us. At the end of our first week, we had a major white glove inspection. When the inspecting NCO (Noncommissioned Officer) could find no infractions in my equipment, he disassembled the brass buckle on my dress belt and discovered a tiny spot of lacquer inside the buckle. Unit protocol required removal of the lacquer coating on issued brass items. After the lacquer vanished, the metal was polished vigorously, resulting in a highly reflective shine. The ensuing demerit caused me to lose my weekend pass. Pamela and my parents came to Columbia that weekend. They only were able to visit with me on post for a few hours on Saturday and a few hours on Sunday. I had to eat my meals in the mess hall while they ate at restaurants in the city. The restrictions made for a bittersweet weekend.

The leadership curriculum stressed two basic responsibilities for a military leader. First, he must accomplish his mission. Second, he must take care of his men. A good leader strikes an appropriate balance between the two duties. In most scenarios, accomplishing the mission but losing most of your men is counterproductive. Conversely, keeping one's men in top-notch shape but never accomplishing a mission is total failure. I learned many other valuable principles during the two-week curriculum. In some cases, the material naturally pertained to military operations. In other cases, the principles were applicable to any field. These lessons served me well during the rest of my military career and in my civilian life afterwards.

The large number of accumulated demerits lowered my class standing at graduation. Consequently I ranked 21st in a class of 39. One soldier dropped out and two failed, leaving 36 in graduating class 47/70 on June 19, 1970.[28] Having completed the course, I received an accelerated promotion to Private E-2.[29] Additionally I was appointed

---

28. AJJ4B-L 19 June 1970.
29. Disposition Form 2496, reference AJJ4B-L, 19 June 1970.

a squad leader for my AIT (Advanced Individual Training) training cycle. A squad leader theoretically commanded an army subunit composed of ten soldiers.[30] My squad at Fort Jackson contained 14 soldiers. The squad leader was responsible for the combat readiness and personal welfare of the men within his squad. He ensured that they received necessary instructions and that they were properly trained to perform their assigned tasks. During AIT, squad leaders wore a black armband with yellow sergeant's stripes on their left sleeve.

ADVANCED INDIVIDUAL TRAINING

At the completion of Leader School, two realities affected my military career. First, an excess of ROTC (Reserve Officer Training Corps) Second Lieutenants led to a cancellation of my application for OCS. ROTC provides financial assistance for college in exchange for military service after graduation. Most ROTC graduates in the class of 1970 had enrolled when public support of the war was still strong and patriotism motivated college students to serve in the armed forces. With combat operations in Vietnam winding down in 1970, the size of the army was shrinking. Therefore West Point and ROTC programs provided sufficient officers for the army's immediate leadership needs.

I wanted to be a paratrooper. Before the cancellation of my OCS

---

30. During the Vietnam era, an infantry company was authorized six officers and 158 enlisted men. It routinely was composed of three rifle platoons and one heavy weapons platoon. A rifle platoon generally consisted of two or more squads. Each squad was divided into two or more fire teams. Training companies sometimes exceeded the authorized strength; whereas in Vietnam infantry companies almost always were markedly understrength.

orders, I had been advised to wait until after completing OCS to apply for the airborne school at Fort Benning. The pay for a second lieutenant in airborne training was much better than what a private earned. I attempted to enroll in airborne school upon completion of AIT. But the army deemed my application as submitted too late. Consequently, I would be assigned to an infantry unit upon completion of AIT. I almost certainly would be going to Vietnam. Ironically, I would spend a majority of my active duty as an infantryman in the famed 101st Airborne Division.

Second, due to a shortage of non-commissioned officers in the training cadre at Fort Jackson, I was assigned greater responsibility as a squad leader in AIT. The added responsibility granted me some special privileges however. Foremost of these was a private room in the company barracks.

The barracks on Tank Hill had been 30-year-old wooden structures. The structures for AIT were new modern brick buildings with eight man rooms for trainees. My smaller room was designed for the unit's permanent cadre. Consequently, not only did it provide me with privacy, it routinely was ignored during barrack inspections. This inattention freed me from strict adherence to the regulation display of clothing and equipment in the barracks. However, as a leader I needed to set an example for the men in my command. Therefore, I was careful never to abuse the situation and routinely kept my quarters in strike military appearance.

Basic training had focused on the fundamentals of being a soldier. Weapons training covered the M-16 rifle, the bayonet, and grenades. AIT consisted of a nine-week course that specifically concentrated on the infantry experience. While conditions in such locations as Europe were not ignored, the focus of the curriculum was designed particularly for combat in Vietnam. We were issued the standard regulation infantry equipment, including the M-1 steel pot helmet, the

M1956 web gear, a bayonet and entrenching tool. We qualified with the M-16A1 assault rifle, the M-14 infantry rifle, the M-79 grenade launcher, the M-60 (7.62mm) and the M-2 (50 cal.) machine guns, and the .45 automatic pistol. I particularly enjoyed firing the M-14. The rifle range employed random popup human torso silhouettes dispersed between 50 and 300 yards. When struck by a bullet the target fell over. For qualifying on the M-14, each soldier was given two 20-round magazines with which to hit 38 such targets. I hit all 38 and had one bullet left. Infantry trainees also learned how to set out mines and dig foxholes.

Additionally we developed familiarity with the enemy's weapons, especially the AK-47 assault rifle. Demonstrations frequently compared the AK-47 with the M-16. These presentations naturally were designed to inspire one's confidence in the American weapon. One particularly impressive comparison involved two metal ammunition cans filled with water. The instructor fired one round from the AK-47 into one can and one round from the M-16 into the other can. The AK-47 round made a nice 7.62 caliber bullet hole in the front where the round hit the can. The exit hole on the backside was identical in size, creating two matching fountains of water streaming out of the can. The M-16 round made a small 5.56 caliber hole in the front. However, the entire backside of the ammo can was a warped, mangled panel with an exit cavity more than an inch in diameter. The lower velocity of the M-16 hunted an unpredictable exit path. Therefore, it inflicted greater internal damage inside the human body.

One day my squad was detailed to clean the grounds around the base main PX. A member of the permanent cadre was placed in command of the detail. I served as his second in command. We stood together on the sidewalk watching the detail work when a female soldier approached. She was in her dress greens and extremely attractive. We were so absorbed in looking at her gorgeous legs that we failed to notice the two silver bars on her shoulders. Consequently, we failed to present the required salute, a remiss for which she straightway chastised us. Her no nonsense attitude distinctly suggested her human vanity lay in military decorum, not sexual charm. Fortunately, for me, the soldier in charge received the brunt of her wrath. Following that incident, I attempted to duplicate the perspective of Private Will Stockdale, Andy Griffin's character in the classic movie *No Time for Sergeants*. "I don't notice whether it's a man or a woman or what. All I see is a captain. That's all," the unsophisticated Stockdale alleged.[31] Then again I must admit that I never achieved such complete gender extinction. As luck would have it, after that incident I rarely saw many female officers.

The sixth week's training schedule in AIT was typical of daily activities and offers some insight into the overall program. That particular week infantry trainees were on duty 58 and a half hours. A majority of these—32 hours—focused on rifle squad tactics. Physical training involved both calisthenics and road marches. Calisthenics developed strength and endurance. Road marches conditioned soldiers to rapid hiking over long distances. These marches traversed various types of terrain during different weather conditions.

We also underwent immunization in preparation for deployment to Southeast Asia.[32] These injections seemed endless. My "International Certificates of Vaccination" booklet lists me having 17 different vaccinations. They usually were given in the field during chow. On one occasion, we lined up and walked through a gauntlet of medics with pneumatic injection guns. If you flinched or they moved, the gun cut a gash. That day we got four or five shots. The sheer volume of inoculations in a limited time made meticulousness injections

---

31. *No Time for Sergeants*, directed by Mervyn LeRoy, Warner Bros. Pictures, 1958.
32. Training Schedule for Co "B" 11th BN for Week 27 Jul thru 1 Aug 1970

impossible. Many soldiers walked out of the sequence with blood streaming down one or both arms.

A majority of men in my squad were in the National Guard. They had secured enlistment in their state units in order to avoid service in Vietnam.[33] Therefore, they frequently failed to take training for combat in Vietnam seriously. By the second or third week, their nonchalant attitude was apparent. My leadership training and previous work experience taught me how to recognize problems, make an accurate estimate of the situation, and then take appropriate action. Informal instruction was one useful method for dealing with some problems. This seemed like one such case to me. So one evening just before lights out, I collected the squad together in one of the barrack's rooms for some conversation. I don't remember much of what I said, but I made it clear that their indifferent approach to training might get me killed. The impact their behavior might have on me had never crossed their minds. After that night, they may not have been the best squad in the company but I would rate their performance as above average…and I was not killed in Vietnam. Looking after the other guy's fate is basic to the infantry experience. In combat, infantrymen usually do whatever is best for their buddies without thought of the personal consequences. Infantry soldiers willingly die in order to save the lives of their comrades.

---

33. The vast majority of US Army units in Vietnam were from the regular army. Only a few National Guard units were activated and deployed to Southeast Asia. These included Kentucky's 2nd Battalion, 138th Artillery; Indiana's Company D (Ranger), 151st Infantry; Hawaii's 29th Infantry Brigade; Kansas' 69th Infantry Brigade, with one infantry battalion from the Iowa Guard; California's 1st Squadron, 18th Armored Cavalry; New Hampshire's 3rd Battalion, 197th Artillery; and numerous engineer, postal, medical and support units.

Transforming teenage boys into effective combat soldiers necessitated psychological changes in their thinking. The army's approach was uncomplicated. Two examples will suffice. During bayonet exercises one hot day, the company fixed bayonets and formed a circle around the instructor. Then for ten or fifteen minutes, we heaved the weapon up and down while yelling at the top of our lungs, "Kill! Kill!" Second, during road marches we usually chanted the drill sergeant's favorite cadence.

I wanna be an airborne ranger
I wanna live a life of danger
I wanna go to Viet Nam
I wanna kill some Viet Cong

The conversion from peaceful lethargy to prepared killer was essential. A day would arrive when it would determine who lived and who died.

In August 1970, a train carrying nerve gas passed through South Carolina.[34] My AIT company was deployed to act as railroad crossing guards during the movement. We rode in a truck convoy to a small town—I don't remember its name—and stayed in the local National Guard Armory. Teams were assigned to every railroad crossing in that part of the state. I however was part of the backup contingent and never left the armory. We were well received in the community during the two days we were in town. People constantly visited the armory. Some ladies even brought us cakes and cookies. A few teenage girls flirted with us from the sidewalk, but I don't think any soldier secured anything more than a girl's first name and some giggles.

---

34. See http://www.epa.gov/region4/foiapgs/readingroom/camp_lejeune/trianado-c21a.pdf (United States Environmental Protection Agency website).

The following week the company held its final field exercise. The training included simulated operations against the Viet Cong. Base cadre dressed in Vietnamese "black pajama" outfits acted as enemy soldiers. The exercise terrain contained a small village of bamboo hooches, hidden bunkers, and other Vietnamese icons.

Graduation followed. Thereafter we could wear the light blue infantry cord on our right shoulder and infantry blue plastic discs behind our hat and collar brass. To date, all of my infantry training had been theoretical. That was soon to change. I left Fort Jackson for ten days leave before reporting to Oakland Army Base for assignment to an unspecified infantry unit in Vietnam.

TEN DAYS LEAVE

For the next 10 days, I tried to forget that I was a soldier. My mind focused on romance instead of war. I was in love with Pamela Coffman and wanted to spend every second I could with her. On Sunday evening, I headed to her house so we could go to church together. For some reason I uncharacteristically was late. I was so absorbed in getting to her house as quickly as possible that I became oblivious to my speed. Then I heard a siren behind me. I glanced at my rearview mirror and saw a flashing red light. I pulled over on the shoulder of the road, rolled down my window, and waited for the Smyrna policeman to cite me.

"Where are you going in such a hurry?" the officer asked.

"Church," I replied, annoyed by the delay he had precipitated.

"If you don't slow down, you're going to arrive dead," the officer retorted.

"I'm already as good as dead," I brashly snapped back without thinking. "I'm going to Vietnam next week."

The smart remark expressed an attitude common in infantrymen during the Vietnam War. It was a mixture of candid fatalism and macabre conceit. It expressed a sense that you can't harm me because I already have been dealt a worst hand. Grunts codified it with the rhetorical question, "What are you going to do to me, make me a grunt and send me to Vietnam?" As future circumstances revealed that things could indeed become worse, the vocal expression became, "F*** it. It don't mean nothin'."

Fortunately, the policeman didn't arrest me. Instead, he calmly asked if I had orders to prove I was going to Vietnam. I reached into my wallet and handed him a folded copy of my orders. He let me off with a warning. The officer started back to his patrol car. Then he turned and said, "Good luck, buddy."

I answered "Thanks" and resumed my drive to Pamela's house at a legal and safe speed.

* * *

Pamela and I had been talking about marriage since January. Originally, we planned to get married as soon as she finished high school. Vietnam changed that. I lobbied for us to marry before I left, but she refused. Pamela was apprehensive of becoming a widow at 16 or 17 years of age. However, she was entirely committed to getting married as soon as I returned home from Vietnam. Even so, she lacked the traditional diamond ring that went with a formal engagement.

On the night we first discussed marriage, Pamela had shown me a picture of the rings she wanted. Just before finishing AIT, I came home on a weekend pass. That Saturday Mother and I went to the jewelry department at Rich's in Cobb Center. There in one of the glass cases was the exact engagement ring and wedding band that Pamela desired. Even with Mother's employee discount, the rings cost more than the army paid me in six months. The engagement ring had to be sent away in order to set the diamond I had selected. Both rings also had to be sized while they were at the jewelers. I paid half of the price that weekend. The rest of the payment was due when I picked the rings up. I was able to complete the purchase during my leave. Now all that remained was to give Pamela the engagement ring.

One evening during my leave, we sat cuddling on a plaid love seat in the den of my parents' house. Pamela wore my high school senior ring on her left ring finger, as was the customary token of adolescent commitment in that era. While we were kissing, I managed to break the rubber band that she used to hold the ring on her hand.

Then I said, "My ring won't fit your finger without a rubber band. Guess I need to replace it." Instead of getting up and going after another rubber band, I pulled out a box that I previously had hidden between the cushion and the couch's arm. I looked into her face and confidently asked, "Will you marry me?"

I don't know what she said. I do know that passionate kissing followed. Whatever the words, they meant, "Yes, I will marry you." That engagement ring has remained on that finger to this very day. I lost the high school ring later somewhere in Vietnam, probably in the South China Sea while I was at China Beach on an in-country R&R (Rest and Recreation vacation).

* * *

I started feeling sick the day before leaving for Oakland. A sinus infection caused a sore throat. I really needed to see a physician. Because I was in the military, that meant going to sick call at nearby Dobbins Air Force Base. I feared doing so might delay my departure and I was determined not to postpone going to Vietnam. The sooner I got "in-country," the sooner I would get back home to Pamela. Therefore,

I flew out of Atlanta for San Francisco on my scheduled flight. My sister Reita was a senior in high school. Reita had a stoic personality and refused to show her emotions publicly. She unashamedly cried aloud in class that day.[35]

---

35. Reita M. King, e-mail message to the author, October 8, 2013.

2. RED WARRIORS

Arrival in Vietnam and Duty with the 4th Infantry Division

OAKLAND, CALIFORNIA (USA) TO LONG BINH (RVN[36])

On September 11, 1970, my flight from Atlanta landed in San Francisco on schedule. I promptly obtained ground transportation across the bay and reported to the Overseas Replacement Station in Oakland, California. I do not recollect much about my brief stretch at Oakland Army Base. I still was very ill but refused to go to sick call. I realized that any postponement to my departure time for Vietnam would delay when I came home. The sooner I left for Vietnam, the sooner I would return home from Vietnam.

In Oakland, soldiers bound for Vietnam were issued five sets of jungle fatigues along with olive drab boxer shorts, undershirts, and handkerchiefs, two pairs of jungle combat boots, and three

---

36. RVN was an acronym for Republic of Viet Nam, the official name for South Vietnam.

olive drab towels. I took off the khaki summer dress uniform I was wearing and put on a set of the jungle fatigues. They were very different from the stateside utility fatigues. The fabric was lighter. The cut was much looser and more comfortable. The jacket was not tucked inside the trousers, but hung loose over the hips. Its front pockets were larger and there were two additional pockets below the waist on its coattail. The pants sported two large cargo pockets on the outside of the thigh. All pockets except the front pants pockets were expandable with a double flap. Overall, jungle fatigues were comfortable and very functional. Small blackened metal pins denoting rank were fastened to the collar. I attached a pin with a single chevron denoting I was a Private (E-2). Upon reaching one's assigned unit, grunts normally removed these pins. Insignias of rank rarely were worn in the field.

Three days after reporting to Oakland, my name appeared on the manifest for transportation to Vietnam. The orders were dated "14 September 1970."[37] This date was of greatest consequence to me. It fixed my DEROS date as "13 September 1971." DEROS was a military acronym for "date of estimated return from overseas." During the Vietnam War, a standard tour of duty in that combat zone for the Army was fixed at one year. DEROS marked the projected date the tour of duty would be finished. Thus soldiers commonly knew from the beginning of their tour exactly what day they would leave Vietnam. For the next twelve months, I always would be conscious that my DEROS date was September 13. Virtually everything I did was measured in relationship to my DEROS date. For instance, shortly after arriving in Vietnam I wrote home in a letter, "*I have 354 days of*

---

37. Headquarters United States Army Personnel Center, Special Orders Number 257, 14 September 1970.

*time, but that's two weeks gone."*³⁸ In a later letter I wrote, *"I'm fine, real fine considering I only have fifty-five days left over here. I'll sure be glad to get back home."*³⁹

When I reached the juncture that only three months were left in-country, I would become a "short-timer."⁴⁰ At that time I would start keeping my own personal "short-timer's" calendar in anticipation of returning to "the World."⁴¹ My calendar was a photocopied drawing of a nude girl. Her body was divided into 91 separate spaces. Each day I colored in a space with a flesh colored marker.⁴² Consciousness of my DEROS date commonly shaped my mental state and behavior during the entire year I served in Vietnam.

Along with the other enlisted replacements, I rode a military bus from Oakland to Travis Air Force Base. The bus meandered through Northern California's rolling hills. This was my first sight of the west coast of the United States. The view was captivating. Golden grass coated numerous slopes. The trip took slightly over an hour. At Travis,

---

38. Letter to my parents while on FSB Augusta, dated 27 Sep 70. All dates on personal correspondence are given in the format in which they were written in the letter.
39. Letter to my father dated July 20, 1971.
40. "Short-timer" was a slang nickname for soldiers whose tour of duty was almost over. Generally the term began to be applicable when the soldier had less than 100 days left in country. The sobriquet "two digit midget" was given to those between 99 and 10 days left until DEROS.
41. "The World" was slang for the United States. It emphasized the stark contrast between life there and life in Vietnam. The United States also was known as "the land of the big PX" because there you could purchase almost anything you desired.
42. My short-timer's calendar was given to me by our company clerk. The last five days were never colored. By that time, I was in transit to my departure location and the reality of leaving Vietnam was overwhelming.

we boarded a chartered Seaboard World Airlines' DC-8 for the flight to South Vietnam. The plane took off shortly before 10:00 o'clock in the morning and flew up the Pacific coast to Anchorage, Alaska.

Four and a half hours later the flight landed in Anchorage. The airplane taxied and parked out on the tarmac. Its fuselage door opened. Metal stairs rolled up to the doorway. Passengers disembarked the airplane while it was refueled. A light snow fell as we walked across the tarmac to the civilian terminal. Our newly issued jungle fatigues offered little insulation against the cold artic air. Fortunately, my health had improved significantly and breathing the cold air felt invigorating.

Upon entering the terminal building, a gigantic stuffed Alaskan Brown Bear immediately caught my attention. The beast stood erect on its hind legs and was at least ten feet tall. There also were two Rainbow Trout 18-20 inches long that impressed me. The trout I caught in North Georgia were considerably smaller. Before leaving the terminal, I purchased a handful of picture postcards of Alaska, intending to use them in jest. I only sent one or two with any message. These included a satirical, "*Having a great time*," or something to that effect. I sent eight of them in a letter to my father without any messages, knowing that he would enjoy the animal photographs on the cards.[43]

Less than an hour later, the airplane was refueled and we resumed our trip. The flight crossed over the rugged snow covered Alaskan mountains. They were both inspiring in their majestic beauty and terrifying in their bleak isolation. Thereafter the airplane proceeded across the Pacific Ocean to Wake Island where it again landed for refueling. The stop at Wake was mundane. The final leg of the journey terminated at Bien Hoa Air Force Base in the Republic of South

---

43. Letter written in the airplane and dated 14 Sep 70.

Vietnam. The entire trip had taken over 24 hours, but it was not without some drama.

As the flight approached Bien Hoa, the aircraft's pilot reported over the intercom that incoming rockets would delay landing temporarily. Whether his announcement was factual or a ruse, I do not know. However, it was a stark reminder that Vietnam was a war zone. The plane circled the airfield briefly and then landed without further incident.

The airplane taxied towards the terminal building and parked. The ground crew rolled out a portable staircase and the flight crew opened the cabin door. An alien environment jolted replacements debarking from the air-conditioned plane. The suffocating warmth of Vietnam's tropical climate sucked the life out of one's lungs. The foul air reeked with the stench of urine and diesel fuel. The heat distorted vision on the scorching tarmac. Breathing tightened as the recent arrivals labored towards the terminal gate. For me it also was a moment of keen introspection. I had landed in a really bad place. I soon would discover a grunt's world was much worse and far more alien. In his book *Nam-Sense* Arthur Wiknik, Jr. wrote:

> "Vietnam was far away from America and we were even farther. Grunts were so detached from everything that it felt as if we were on a planet in outer space while everyone forgot where we were."[44]

Wiknik nailed it. Major news events occurred while I was in the bush and I didn't learn about them until years later.

By the time the habitual military procedures were complete and

---

44. Arthur Wiknik, Jr., *Nam-Sense: Surviving Vietnam with the 101st Airborne Division* (Philadelphia: Casemate, 2005), p. 29.

all baggage had been claimed, the sun had started setting on the western horizon. Along with the other replacements, I tossed my duffle bag into a flatbed truck and quickly filed into one of the olive drab buses parked in line nearby. I sat down. *A window seat! I'll be able to see what Vietnam really looks like.* I abruptly noticed wire metal grates covered the windows of the bus. *Not a good omen.*

My prospect of sightseeing also was diminished by the lateness of the hour. Darkness already hid the world outside as the buses pulled out of Bien Hoa. The road to the Transient Detachment at Long Binh was a dirty two-lane asphalt thoroughfare lined with closely packed one and two story structures set approximately ten to twenty feet off the road. Only a small number of the structures displayed Asian architectural décor. Most were just plain rectangular boxes. These unpretentious edifices were constructed with a variety of materials, ranging from wood or plastered concrete to corrugated tin or cardboard. The sounds of exotic music drifted from a few of the buildings. The numerous lighted windows in these structures provided brief glimpses of the mysterious, indigenous culture. As the highway approached Long Binh, the shoulder opened into fields and the buildings were spaced further apart. The pavement widened and a dashed white centerline separated the two lanes.

At Long Binh, the highway entered a fenced compound. A large white sign spanned the entrance. "90TH REPLACEMENT" over the word "BATTALION" was printed in large black lettering with gray shadows. Unit insignias were painted on either side, the blue and gold star of the battalion on the left and the shield of the USARV[45] on the right. Inside the barbed wire stood numerous unpainted one and two story wooden structures. We filed off the buses and climbed a dumpy embankment onto a red dirt parade ground. Opposite the

---

45. United States Army, Republic of Vietnam.

parking lot was a short tower with a speaker. A white plywood signboard served as its front. "WELCOME TO VIETNAM" was emblazoned at the top of this sign. Here replacements were given instructions and assigned to a barracks for their time in the replacement station. This structure made the World War II basic training barracks at Fort Jackson seem like a five-star hotel. The floor of the Vietnam structure was unpainted concrete. Walls were wood boards over wire screening on 2x4 wood studs. The wall boards were slanted to allow ventilation. The roof was corrugated tin. Random sandbags were scattered over the metal.

During the next few days all replacements processed "in-country" before they were assigned to their new unit. The usual army hassle filled meaningless time but some endeavors conceived more ominous tones. Roll calls were scheduled periodically throughout the day. Unit assignments and transportation rosters were announced at these formations. Work details also were formed. In many ways the routine was no different from that experienced stateside.

However, there were countless reminders that this was not stateside. American currency was exchanged for Military Payment Certificates or MPC. Soldiers branded it "funny money" because of its size and appearance. MPC was brightly colored paper money in denominations from five cents to twenty dollars. It was the legal tender on all American bases in Vietnam during the war. It was illegal for unauthorized individuals to possess MPC. Nevertheless, a significant black market traded in these notes and local Vietnamese merchants usually accepted MPC.

Therefore, periodically MPC notes were changed. In early October—while I was at Camp Radcliff—MPC changed from Series 681 to Series 692. The new currency was decorated with buffalos, eagles, and Native Americans, whereas the Series 681 were printed with military vignettes such as fighter jets or nuclear submarines. The camp was locked down without warning and all notes were exchanged in a single day. Subsequently the old Series 681 notes became worthless. Rumors claimed several civilians in the neighboring town of An Khe committed suicide the day after the exchange. On the other hand, "Boom-boom" girls steadily slipped on base to start recouping their losses.[46] Because of the money's novelty, I mailed a five cent note and a one dollar bill from the Series 681 home from Long Binh as souvenirs. However, I did not keep any Series 692 notes. By the time these notes were issued I had become so accustomed to using MPC as money that wasting them as souvenirs no longer occurred to me. Instead MPC's monetary value was all that mattered.

The most portentous clue that this was not stateside duty lurked in the paperwork replacements filled out. One form designated beneficiaries for government life insurance policies. Other forms pertained to notification of next of kin in case of injury or death. In cases of minor wounds, soldiers were permitted an option of not having their families notified. I sat at a wooden picnic table outside as I completed these forms. Contemplating my own potential demise truly was surreal. Death and injury were not mere theoretical risks. They were real possibilities and the statistical odds were not encouraging. Over two million Americans served in Vietnam during the war. According to military records 58,220 Americans died in Vietnam.[47] Another 153,303 were wounded in action and 10,173 more were captured or missing in action.[48] I chose the option of not having my family notified unless the injury was fatal or serious. I signed the paperwork

---

46. "Boom-boom girl" was slang for prostitute.
47. National Archives, Military Records website (https://www.archives.gov/research/military/vietnam-war/casualty-statistics). Accessed March 9, 2018.
48. *The Encyclopedia of the Vietnam War*, ed. Spencer C. Tucker, 2000 ed., s.v. "Casualties" by Spencer C. Tucker, 64.

and sat there staring at this bizarre and unfamiliar world in which I found myself. In time I would grow accustomed to young men dying or being maimed by other young men. Death had been infrequent in my past. An older cousin died in a hunting accident when I was 13. My paternal grandfather died when I was 18, but he was 83 at the time. Basic training so occupied my existence that I had little time to think about Tommy's death in Cambodia. Now the thought of dying was inescapable.

This realm of human violence would soon become normal and a life of peaceful pursuits would seem strange and alien. However, that day at Long Binh, everything still looked incomprehensible and bewildering. Surely, it was in some dreadful nightmare. Surely, I would awaken from this bizarre dream and discover my customary life back in the real world. However, "surely" never came. Vietnam now was reality… my reality for the coming year.

CAMP RADCLIFF AND
FIRE SUPPORT BASE AUGUSTA

The second morning at Long Binh, I stood in one of the endless formations. Then it happened. I heard my name called. "Matthews, Edwin L." I listened carefully to what was said. I had been assigned to the Fourth Infantry Division. All troops assigned to that division reported to an open shed with a designated number painted on a white placard at one end. From Long Binh I was sent to the 4th Replacement Depot at An Khe in the Central Highlands for assignment to one of the division's line companies. Up to this point everything had been a totally novel occurrence. I never had faced anything like it. I responded like a "deer in the headlights." Now I had to shake off my lethargy and grasp every clue to this peculiar world in which I found myself. I likely would not survive otherwise.

The 4th Infantry Division was organized in 1917 for service in World War I. The division's motto is "Steadfast and Loyal," but it was nicknamed "the Ivy Division." The moniker was a word play on "IV," the Roman numeral for four. The division went to Europe as part of the AEF (American Expeditionary Force) and fought at Aisne-Marne, St. Mihiel, and Meuse-Argonne.

During World War II, the 4th Infantry Division landed on Utah Beach on June 6, 1944, in the first wave of D-Day. It fought its way out of Normandy and became the first American division to reach Paris. Later the Ivy Division earned the distinction of being the first Allied division to cross the Siegfried Line and enter Germany, on 11 September 1944.

The lead elements of the 4th Infantry Division arrived in the Republic of Vietnam on 6 August 1966. It was based in the strategic Central Highlands. Located in west-central South Vietnam, elevations in the region range from 3,000 feet above sea level to over 8,000 feet.[49] During the four plus years that the division was deployed to Vietnam, it operated from the Cambodian border near Dak To and Pleiku to Qui Nhon and Tuy Hoa on the coast of the South China Sea.

I flew from Bien Hoa to An Khe in a C-7 Caribou cargo plane. The C-7 [CV-2] originally was designed as an Army all-weather transport. The twin-engine prop aircraft was relatively small for a cargo plane, but its short takeoff and landing capability suited it for heavy service in Vietnam.[50] The veteran airplane in which I flew had

---

49. *The Encyclopedia of the Vietnam War*, ed. Spencer C. Tucker, 2000 ed., s.v. "Central Highlands" by Claude R. Sasso, 66.
50. Stanton, Vietnam Order of Battle, 284-85.

several makeshift repairs. One hole had been patched with aluminum from a soda can and the Coca-Cola™ logo was clearly visible. In another place, wire from a coat hanger replaced two missing rivets. Crammed into the cargo compartment of this small aircraft, I saw very little of the terrain below. Nevertheless, the flight was brief and uneventful.

The American combat base at An Khe was Camp Radcliff. Its most distinctive feature was the inclusion of Hon Cong Mountain (also known as Hong Kong Mountain) within the base's defensive perimeter. Whereas most of the base was located on relatively flat terrain, this high mountain jutted high above the surrounding countryside.[51]

At Camp Radcliff, replacements received division orientation and additional training in jungle warfare. The week-long instruction time also allowed replacements to adapt to the tropic climate. I pulled my first guard duty in Vietnam during this training period. I was terrified. Hunched in a sandbag bunker staring into the darkness, one's mind easily can see battalions of enemy soldiers creeping towards you. By the time I completed this phase of my tour, my face was sunburned,[52] but my arms and neck had tanned nicely. I cannot account for this incongruity other than the tanned areas must have been exposed more to the sun previously.

My first tangible exposure to combat occurred during this training

---

51. Albert N. Garland, ed., *Infantry in Vietnam* (Nashville: The Battery Press, 1982), p. 254, Map 70; Michael P. Kelley, *Where We Were in Vietnam: A Comprehensive Guide to the Firebases, Military Installations and Naval Vessels of the Vietnam War* (Central Point: Hellgate Press, 2002), pp. 5-246, 5-433.

52. In recent years, numerous skin cancers (basal cell carcinoma) have been surgically removed from my face. According to the dermatologist, these cancers likely have their origin in this sunburn.

period at Radcliff. The pilot's announcement about incoming at Bien Hoa had been little more than hearsay. Other than a few minutes delay I had no concrete evidence it actually occurred. For all I know the alarm may have been the pilot's macabre sense of humor. This time I heard the explosions. On the night of September 21-22, the enemy fired four or five mortar rounds at the base. An alarm sounded and we raced to nearby bunkers. The next morning I wrote my parents, "*They didn't hit nothing and only suceeded* [sic.] *in waking me up about thirty minutes early.*"[53]

I was an active evangelical Christian when I entered the army. The tour in Vietnam solidified my Christian beliefs. I was able to share my Christian faith with other soldiers on several occasions. The first of these was while I was in the Replacement Depot. During the months ahead there were times my behavior disappointed Christ, but my personal relationship with Him sustained me through this difficult era of my life.

By 1970, Vietnamization dominated the United States strategy in Southeast Asia. Vietnamization was a coined term that denoted the policy of progressively shifting the primary burden of the war from the United States military to the South Vietnamese military. American troop levels were being cut dramatically as South Vietnam assumed more responsibilities in the war. Whole divisions began to depart Vietnam for redeployment back in the United States. In April 1970, the First Infantry Division returned home.[54] That same month President Richard Nixon announced the withdrawal of an additional 100,000 troops by December, at which time total American strength in country would fall to 184,000.[55] Consequently rumors abounded in Vietnam about who

---

53. Letter to my parents dated "22 Sep 70."
54. Stanton, *Vietnam Order of Battle*, p. 74.
55. Jacob Neufeild, "Disengagement Abroad—Disenchantment at Home," *The*

would go home and when. The 4th Division expected to redeploy shortly and I hoped to return home to "the World" with the division.

Popular opinion today often mistakenly assumes American units experienced little combat during this period of the war. Engagements tended to be on a smaller scale and their frequency dependent upon the geographical region where a unit was deployed. Along the Cambodian and Laotian borders or near the DMZ, contact with the enemy occurred regularly. When the two sides encountered each other, soldiers died. While Vietnamization was a factor in the change, another reason was due to an alteration in the tactics of the North Vietnamese. The NVA (North Vietnamese Army) rarely deployed large units within South Vietnam during this period of the war. They had suffered too heavily in pitched battles like Ia Drang and Dong Ap Bia (Hamburger Hill). Instead, they utilized smaller, more elusive units. Consequently the American army generally employed platoon or company sized units to search out and destroy the enemy rather than whole battalions or regiments. While the smaller unit size yielded fewer numerical casualties in a given engagement, these firefights were no less intense or bloody.

I was promoted to Private First Class on September 23 and assigned to Alpha Company (Company A), 1st Battalion, 12th Infantry Regiment,. The regiment was known as "the Red Warriors."[56] That same day Alpha Company sustained its last combat deaths in Vietnam. Hiram Edenton Jr. and Sergeant Robert Riveria were killed by enemy indirect fire.[57] Hiram Eurias Edenton, Jr., from Spotsylvania,

---

Vietnam War: The Illustrated History of the Conflict in Southeast Asia, Roy Bonds, ed. (New York: Crown Publishers, 1979), 210, 217.

56. Headquarters 4th Infantry Division, Special Orders Number 266, Extract 2. TC 215, and Extract 23. TC 422, 23 September 1970.

57. 1/12th Red Warriors Vietnam Association website (http://redwarriors.us/Histo-

Virginia, died two days after his twentieth birthday. Robert Charles Rivera, from Honolulu, Hawaii, was only 19 years of age.[58]

Except for the set I was wearing, at Radcliff I turned in all of the tropical uniforms that I had been issued in Oakland. I kept one towel. It would be several weeks before I changed clothes again. When I did, the clean clothing would consist of older items from the division laundry instead of the new uniforms I had been issued. Thereafter I rarely wore a jacket with my own name embroidered above the pocket.

I reported to the company orderly room at Radcliff. At the time, Alpha Company was operating out of Fire Support Base Augusta. Before joining them in the field, I was issued the necessary field gear. The next morning I flew by helicopter to Augusta. I had flown in a Huey only once during AIT. On that occasion I sat strapped into the helicopter's seat. On this helicopter the seating had been removed. Boxes of supplies were stacked in the interior of the aircraft bay. The four passengers sat on the floor with their legs dangling in space. My first flight had been over South Carolina. This flight was over South Vietnam. Curiosity and trepidation fought for control of my thoughts.

Soon FSB Augusta came into view. The site was a barren mountain peak. All vegetation on the site had been eradicated long ago, leaving only hues of red dirt that covered everything in layers of fine dust. The firebase extended down onto a saddle. A small knoll extended off one side of the saddle. Concertina wire surrounded the installation. Bunkers had been constructed from sandbags, wooden mortar round crates, and metal stakes. Their size varied. A ring of small bunkers formed the firebase's perimeter. Larger bunkers were

---

ry1970.htm) Accessed June 7, 2017.

58. The Virtual Wall ® Vietnam Veterans Memorial (www.virtualwall.org). Accessed June 7, 2017.

crammed into the firebase's interior. Radio antennas reached skyward. Howitzer barrels pointed to distant peaks and valleys. Saw grass and scrub bushes covered the slopes outside the wire. As one descended into the valley, these gave way to dark green jungle.

A thick cloud of red dust swirled around the landing pad as the helicopter landed. The passengers jumped down and ran to clear the overhead rotor blades. Two of them were veterans returning to Augusta. Neither was in Alpha Company. Jimmy Todd and I were replacements. The two veterans immediately scurried out of sight. At the same time, a small detail of men rushed up and began unloading the supplies. They slung boxes and crates out with robotic speed. As the last crate hit the ground, the helicopter slowly lifted up. Then it bolted and soon disappeared from sight. Several additional soldiers joined the detail and hastily dispersed the recently procured supplies to various parts of the firebase. Jimmy and I stood bewildered, not knowing where to go or what to do. The two of us stood covered in a fresh coat of red dust, looking like lost sheep.

Finally a lone individual came over and inspected us. He was stripped to the waist. Exposure to the sun had tanned his muscular body golden brown. The distinctive red earth impregnated his clothing, turning its olive green to a pale shade of sepia. The dirt caked his bleached hair and dyed it the same color. His stare sent chills up my spine. Finally he asked, "F-N-Gs?"[59]

"Huh?" I responded.

"FNG? What's a FNG?" Jimmy inquired.

"F***ing New Guy, s*** head," the stranger whispered curtly, as if he spoke only to himself. Then he spoke in an audible voice. "You guys replacements?"

---

59. FNG was an abbreviation for the colloquial expression "F***ing New Guy." In the 101st ABN Division "Cherry" was the preferred slang term of replacements.

"Yeah," we replied meekly.

"Red leg or grunt?"

"Huh?"

"Are you Arty or Infantry, dumb a**?"

"Infantry, Company A, First of the Twelfth Infantry," I answered.

The stranger pointed to a bunker below the landing pad and said brusquely, "The grunts' CP is down there."

I didn't yet know exactly what a CP was, but I was pretty sure the stranger's directions meant for us to go there. Jimmy and I walked down to the stipulated bunker. From our vantage point the bunker appeared only to be a raised square of gray-green sandbags coated in the universal red dust. Two soldiers squatted on its top peering out towards the tree line beyond the cleared ground. They were talking but we could not make out their conversation. Although also covered in the collective red dust, their uniforms still retained an olive drab color. One was bareheaded. The other wore a disheveled boonie cap. Several rucksacks and helmets were scattered around the backside of the bunker. Whereas Jimmy and I clutched our M-16s so tightly our knuckles turned white, their weapons and ammunition lay casually beside them. Hearing us approach, the two grunts turned and gave us the same visual interrogation as the soldier on the landing pad. After what seemed to be forever, the bareheaded soldier lifted a green plastic canteen to his lips and took a swallow.

The grunt in the boonie cap inquired, "Whatcha want?"

"We're looking for Company A, First of the Twelfth Infantry," I replied.

The grunt in the boonie cap glanced at his companion and in a low apathetic voice said, "New guys."

His companion muttered under his breath, "FNGs."

The grunt in the boonie cap continued his evaluation of us and eventually said tersely, "This is Alpha Company."

Despite his initial detachment, the grunt in the boonie cap proved very helpful. He got us checked in with the ranking NCO and then took us to the firebase perimeter where other members of the company waited to be reunited with their comrades. They soon put us at ease — at least as much as possible in our circumstances — and helped us get situated.

Augusta was a semi-fixed artillery position for Battery B, 4th Battalion, 42nd Field Artillery. Their M102 105mm howitzers provided direct artillery support for "Red Warrior" infantry units in the area. This was my first opportunity to observe an artillery battery up close. Around artillery emplacements were red and white poles used in sighting the gun and piles of expended brass shell casings. The vast majority of the gunners were stripped to the waist. A few wore army green t-shirts. Jackets only were worn when not firing the guns. The concussion created when the howitzers fired was impressive on a newbie grunt. Later that evening the awe turned to frustration. Charlie (American slang for the Viet Cong) was most active at night. Consequently the battery received constant requests for fire missions. The guns' loud firing made sleep impossible. Sleep deprivation soon became my constant companion and fatigue from the lack of slumber my normality.

However, I had not been sent to Augusta as a tourist. I was a grunt and had a job to do. Grunt was a popular slang name for an infantryman in Vietnam. Ground pounder was another slang term. Whatever slang nomenclature might be employed, it denoted a combat infantry soldier. The primary role for grunts on Augusta was providing security for the artillery. Accordingly, we occupied the small bunkers around the perimeter on this isolated hilltop. Outside the perimeter, the coils of strung concertina wire were three or more rows deep. Old tin cans containing a few rocks hung suspended from the wire. These cans provided a primitive but effective early warning signal. The jungle around the hill had been cut back a hundred yards or more. The cleared ground provided unobstructed fields of fire for the grunts inside the bunkers.

Each bunker controlled several M18A1 Claymores. These grey-green antipersonnel mines were convex shaped. Each mine was approximately ten inches wide, four inches high, and an inch and half thick. It was set up on the surface of the ground by inserting a pair of folding bipod legs into the soil. A Claymore threw 700 steel balls ahead in an arc of approximately 60 degrees when detonated. Detonation was produced by an electrical charge from a small hand clicker. The clicker was attached to the mine by a 100 feet spool of wire. When squeezed, the clicker sent a three-volt charge to the detonating cap, igniting a half pound of C-4 plastic explosive in the Claymore. The front of the mine was embossed with the words "FRONT" above "TOWARDS ENEMY." This always struck me as humorous. Did the government think we were stupid? Of course the front faced the enemy. Guard duty took on new intensity for me in as much as I now was at a more vulnerable location.

In addition to guard duty, grunts on a firebase performed other menial labor related to the security of the base. They filled sandbags with dirt and used the bags to strengthen the defenses. New rolls of concertina wire were installed. New growth in the fire zone was chopped down in order to deny the enemy any cover if they attacked the base.

The veterans on Augusta warned us new guys not to leave any skin exposed if we slept inside the bunkers. Large rats inhabited them. The rats were fearless and would bite human flesh. Rats were the one species of wildlife not harmed by the war. Instead their population proliferated, especially on American firebases. So to my consternation the situation presented me with a dubious dilemma. Did I seek shelter from Charlie and fight the rats? Or did I seek shelter from the rats and expose myself to Charlie? Rather than wrap up in the hot climate, I decided to relinquish inside the bunker to the rats and sleep outdoors behind the bunker. Actually, I listened to the veterans and watched to see what they did. Then I did the same. Listening to the men with actual experience often was critical to survival in Vietnam. Men with rank—especially green second lieutenants—often made rookie mistakes and people died. Smart leaders listened to the advice of veteran grunts and learned about the real situation on the ground.[60] Thereafter they were better equipped to accomplish the mission while also caring for their men.

A couple of days before I arrived at Augusta, the enemy had mortared the firebase heavily. One man was killed and 17 others wounded. Much of the damage occurred in the mess area. As a result, the mess area was closed until an underground facility could be built. Accordingly I was introduced to what would become my standard diet for the next twelve months, C–Rations. I had eaten a few C-Ration meals at Fort Jackson. At the time they were an intriguing novelty. Now they were the only food available.

---

60. On March 29, 2018, Auburn High School (Auburn, Alabama) sponsored an East Alabama Vietnam Veterans Welcome Home Ceremony. Speakers included Medal of Honor recipients CSM Bennie Adkins and Maj. Gen. James Livingston, and noted Vietnam War reporter and author Joe Galloway. I enjoyed the privilege of having the invocation. During the pre-event meal, we were discussing the differences in the war at various times. Galloway observed, "Vietnam was not one ten-year-long war, but ten one-year-long wars." In reality the situation was even far more complex. Geography and civilian population were major factors. The rugged mountains east of the A Shau differed immensely from the flooded rice paddies of the Mekong Delta. The tactics and organization of the North Vietnamese Army operating in the former likewise differed greatly from the Viet Cong operating in the latter.

C–Rations became the standard field ration for the Army shortly after World War II. Each meal consisted of canned food and an accessory package containing condiments. In 1958, C–Rations were replaced by Meals Combat Individual. These meals were very similar to C–Rations and so troops continued to refer to them as C–Rations. They came packaged in a corrugated cardboard case containing 12 cardboard cartons. Each carton contained one meal packaged in cans and one brown foil package. The cans contained the entrée, fruit, dessert, and crackers and spreads such as peanut butter, cheese, and jam. Some units also contained a cocoa beverage powder. These cocoa packets were especially prized in the monsoon season when a cup of hot cocoa provided a warm boost from the cold, wet misery. The foil package contained salt and pepper packets, sugar, instant coffee, non-dairy creamer, 2 pieces of chewing gum, a packet of toilet paper, small packets of cigarettes, and matches.

The number of meals issued to the individual soldier depended upon the logistics of resupply. Theoretically an infantry unit was resupplied on average every four to five days. Hence each individual soldier commonly received an entire case. Grunts immediately busted open their cases and started haggling with comrades to trade for a favorite entrée.

A person almost could determine how long a soldier had been in-country by the C–Ration meals he ate. Early in one's tour the only entrées that seemed palatable were chicken and noodle soup or beans and wieners. In a letter to my parents just before Thanksgiving 1970 I wrote:

Well I guess I better fix me some chow. I think I'll have pork and beans [beans with frankfurters] *and peaches. Delicous* [sic.], *huh? It's food.*[61]

In time you burned out your taste for these meals and moved on

---

61. Letter to my parents dated "24 Nov 70."

to beef steak or ham slices. Eventually, as your DEROS got near, you craved such delicacies as ham and lima beans (better known as ham and motherf***ers). The latter came in the older issues from the late 50s and early 60s and therefore they were getting rare by 1970. Even so, my comment that "it's food" epitomized the soldier's real critique of C–Rations.

Not many people ever developed a taste for ham and eggs. I recall one occasion in which we had been without food for a day and a half. Someone had a can of ham and eggs in their rucksack and offered to share it. We unanimously agreed to wait until the next day, hoping to be resupplied in the meantime. Rations, in fact, were delivered later that day and we did not have to eat ten-year-old canned ham and eggs.

After completing their trades, grunts customarily set aside the items they wanted for their next meal. There is an old infantry dictum that says food in the belly is easier to carry than food on the back. Consequently the day that the resupply bird dropped fresh C–Rations, grunts ate their largest meal. The rest of his issue was removed from the cartons and dumped into his rucksack.

Alpha Company arrived at Augusta a few days after me. They looked exhausted as they humped up the steep slope below Augusta. The soldiers in the column were spread out in six feet intervals. Their uniforms were filthy and the men reeked with pungent body odor. The heavy loads on their backs slowed their climb up the steep slope, giving me a longer opportunity to observe my future companions and imagine my own fate. The look on their haggard faces suggested the primitive firebase was the grandest mansion in the world. As each man trudged through the barbed wire into the relative safety of the fire base they briefly looked with disdain at the two new replacements in their new green jungle fatigues and quickly moved on to the business of survival. Equipment was checked, weapons cleaned, and duty assignments made.

Later that day I was assigned to the first squad of third platoon. My squad leader was a sergeant from West Virginia called "Teddy Bear." Nicknames were more common than real names and so the true identity of many men with whom I served has been lost to me. Teddy Bear had an oval face with sandy brown hair. His cheeks and nose were notably browner that the rest of his pale skin. A thin blond mustache covered his upper lip. Others in my squad were Doc Kraft (the medic), Flash, Grizzle, Hobson, Keith, and Monty.[62] Jimmy Todd also was put in the same squad. We had been together since Long Binh. His presence made life in this new world a little more tolerable. Although I never went into battle with these guys, they supplied me with valuable insight into surviving as a grunt in Vietnam. Later when I experienced combat for the first time, I was much better prepared than most replacements.

Teddy Bear made me an assistant gunner for the squad's M-60 machine gun. The M-60 was a 7.62mm air-cooled weapon that fires 600 rounds a minute. It weighs approximately 23 pounds. It officially was a crew served weapon but could be operated by a single gunner. Grizzle carried this weapon. Grizzle was a few inches shorter than me but with a burlier build. The sun had tanned his skin earth brown and a dark brown tuff of hair stood up on top of his head. His mouth only hinted at smiling. A pair of dark brown eyes stared into one's very soul one moment, but at other times they appeared to gaze at some distant unknown vista. One could not really determine exactly what he was thinking. He had seen too much of the war. Since I was

---

62. A search on the 1/12th Red Warriors Vietnam Association website (http://redwarriors.us/default.htm) , accessed June 29, 2017, produced no results for any of these names.

his assistant gunner, he particularly mentored me about life in the bush. The assistant gunner's role primarily was carrying extra ammunition and helping feed ammunition belts into the gun. This gun was so vital to an infantry squad that every man in the squad carried at least one belt of its ammunition. Some photographs from the period show grunts draped with belts like the stereotypical Mexican bandit in an old western movie.

Two days after arriving at FSB Augusta, Alpha Company returned to Camp Radcliff for a stand-down.

* * *

Stand-downs were a vital component of an infantry unit's routine in Vietnam. After an extended period in the field, time was needed to refit and relax. At the time, I could not appreciate the significance of stand-downs. Only later, when I served in Company C, 3rd Battalion, 187th Infantry Regiment (Airmobile) did I grasp how important stand-down was to the sanity and combat effectiveness of a grunt. During 1970-1971, Company C customarily spent approximately 45 to 60 days in the bush. Then it returned to base camp for a three to five days stand-down. A typical stand-down began with dropping rucksacks near the helipad and turning in all weapons. Weapons and ammunition were stored in the company armory. This was a locked metal CONEX[63] adjacent to the battalion helicopter landing pad. From the CONEX weary soldiers proceeded to a succession of olive drab canvas tents. We disrobed in the first tent, tossing our filthy clothing into a large pile.

The second tent housed hot showers. After washing away the dirt

---

63. A CONEX was an intermodal container, a standardized reusable steel box that is utilized for shipping and storing freight.

and grime, we proceeded to another tent where a medical officer gave each man a cursory checkup and prescribed appropriate treatment. Once cleared medically, we moved into the last tent. Inside, lying on the ground were piles of clothing. One pile had pants; one had jackets; one had socks; and one had t-shirts (undershorts were never worn because of chafing in the tropical climate). We grabbed one item from each pile and hoped it would come close to fitting. When an item did not fit, one simply tossed it back into the pile and grabbed another one of the same item. This last tent was adjacent to the battalion mess hall. After exiting the tent we were fed a hot meal. Finally we retrieved our rucksacks and walked to our platoon barracks, known as hooches.

During this time in the rear, an infantry company repaired or replaced damaged equipment. Men rested weary bodies and enjoyed various forms of recreation such as USO shows, games, and movies. The unit also participated in training exercises.

* * *

Therefore, my weary comrades in Company A, 1st Battalion, 12th Infantry Regiment, began this stand-down, getting their gear ready for returning to the bush. However, unknown to us at the time, this was to be Alpha Company's final stand-down. The Fourth Division was redeploying to Fort Carson in Colorado. Although the division would not depart for Fort Carson until December 7, Alpha Company ceased all combat operations and began to disperse when it reached Camp Radcliff in October.

Every replacement assigned to Fourth Infantry Division was categorized as a FNG. Frequently the veterans considered the life of an FNG as expendable. However, integrating replacements into the unit became critical for the cohesion of the squad in the field. This was as much a component of stand-down as replacing broken equipment. I was fortunate in that I had an extended period of quiet time for the squad veterans to tutor me. Some supply clerk had saddled me with an obsolete and dysfunctional World War II era plywood back pack rather than the standard Tropical rucksack. This frame had been designed to carry heavy loads, not infantry field equipment. It consisted of solid wooden frame with sides that bent back 90º. A canvas sheet was attached tightly to both sides with laces. An army field pack was affixed to the outside of the board.

The standardized Tropical rucksack was a nylon pack with three outside pockets. There were straps on the sides and bottom for tying additional equipment. The pack was mounted on a spring steel frame. A quick-release buckle was located on the right shoulder strap. Teddy Bear got me the proper rucksack and showed me how to insert cardboard from a C–Ration case into the frame for better support. Wearing an olive drab towel over the shoulders under the straps provided additional padding and absorbed sweat. By nightfall, everyone's gear, including mine, was ready for field duty. C-Rations, extra ammo, grenades, extra clothing (one pair of socks and a jungle sweater), and perhaps a two quart canteen were packed inside the rucksack. Four or more one-quart canteens, smoke grenades, and an entrenching tool were attached to the outside. A poncho and poncho liner were rolled up and stowed on the frame below the pack. An empty M-60 ammo can was used to store personal items such as stationary and envelopes, wallet and pictures, and camera. The can either was placed inside the pack or tied to the poncho roll. On average, a grunt's rucksack weighed at least 60-75 pounds.

The next few days at Camp Radcliff were spent playing cards and softball, going to the PX, and other casual activities. The veterans enjoyed the daily hot meals, but I was too new in-country to appreciate them. Military duties included guard duty and various work details. I was intrigued by the footbridge soldiers had to cross to get to the PX from our hooch. It was made from PSP planking, the pierced (or perforated) steel slats that engineers use to lay temporary runways. The planks were 10 feet long and 15 inches wide with three rows of holes punched lengthwise.[64] The bridge was a favorite photography spot for the troops.

Typical hooches (barracks) in Vietnam were wood frame structures erected on concrete slab or wood plywood floors. Two-by-four studs were set 24 inches apart. Wood siding covered the lower portion of the side walls with wire screening or clear plastic sheeting above the siding. At Radcliff the floor was concrete. The siding was lapped boards over plywood. The interior walls were painted ochre; the studs were an orange tinted red. The exterior was unpainted.

Roofs universally were corrugated metal with random sandbags laid on top. The weight of the sandbags was necessary to prevent helicopter rotor wash from ripping off the tin sheets. At Radcliff, a row of 55-gallon drums filled with earth formed a low wall that surrounded the exterior of the building for protection against enemy fire. In most cases, the only furnishings inside were folding army cots.

Water typically was dispensed in base camps and on fire support bases from the two-wheeled army tanks known as water buffalos. In addition to drinking water, these trailers supplied water for personal hygiene. A typical shower contained a water tower and a small one-room structure. Water was pumped from the water buffalo into the water reservoir on top of the tower. Gravity then fed it through a heating element into showerheads inside the structure.

Two types of latrine facilities were available for the troops to relieve themselves. Urinals, known as "piss tubes," consisted of a small metal pipe stuck directly into the ground. A wire screen covered the pipe's exposed end. The only concession to privacy was a partition

---

64. etp Mickenautsch website (http://psp-marsden-matting.com/psp-history.html). Accessed June 29, 2017.

fabricated by cutting a 55-gallon drum's wall in half vertically and attaching it to two poles that had been driven into the ground on either side of the pipe. This makeshift barrier screened a man's exposed penis, but not much else.

The other latrine facility was a large outhouse. These generally contained a plywood bench with two to four holes. These latrines commonly were fully enclosed structures, although one or more walls might have an opening covered with clear plastic sheets that functioned as a window. Underneath each commode hole was a tub made by cutting a 55-gallon drum in half horizontally. These tubs could be accessed through a lift up door on one exterior wall.

These outhouses unquestionably generated the absolute worst chore for a grunt in Vietnam. I was introduced to this putrescent procedure during this stand-down at Camp Radcliff. As the newest member of the squad, I naturally was assigned to the first "s*** burning" detail that week. Grunts on this detail were given a pair of work gloves, a jerry can of diesel fuel, a metal hook, a wooden stick, and matches. The hook was used to extract the half-drum tub from under the outhouse and drag it out into the open. Approximately a third of the tub already was filled with human feces when the tub was removed from under the toilet. An equal amount of diesel fuel then was poured into the tub. Next the excrement and fuel were mixed by stirring them with the stick. A lit match was tossed into the drum creating a hot fire and a disgusting odor so foul that only those whose nostrils have inhaled the evil concoction can imagine its putrid smell. When the drum's contents burned away, the tub was returned to underneath the outhouse. Thereafter the detail was released and its members usually headed straight for a shower—if showers were available.

QUI NHON

As the short stand-down came to an end, we learned officially that the 4th Infantry Division was returning to CONUS.[65] The initial elation that the division was going back stateside was tempered by reality. Sadly for most grunts in the company, the rumors that the division would redeploy as a whole unit were inaccurate. We would not return to "the world" with the division. Except for a few short-timers, the company's personnel were reassigned to other combat units throughout Vietnam.

After a majority of the company shipped out, I remained at Radcliff another week as part of a company cleanup detail. We packed up equipment for shipment elsewhere and disposed of items not listed in company paperwork. According to army logic, if paperwork authorizing the unit to possess an item does not exist, the item does not exist. In one incident, a case of brand new machetes could not be accounted for through official records. Therefore, the detail destroyed all of the machetes with thermite-filled incendiary grenades. The grenades burned for about 40 seconds at a temperature of 4,300 degrees Fahrenheit.[66] In less than a minute, the machetes burned into useless scrap metal.

After completing my duties on the cleanup detail at Camp Radcliff, I traveled to the Army Depot at Qui Nhon for final processing out of the 4th Infantry Division. I rode to Qui Nhon in a truck convoy down QL 19 (QL abbreviated Quoc-Lo, Vietnamese for National Highway). QL 19 was an old colonial road that connected the port of Qui Nhon to the Highlands capital of Pleiku. Between the two

---

65. CONUS was the military acronym for Continental United States.

66. *Encyclopedia of the Vietnam War*, 2000 ed., s.v. "Grenades" by Robert J. Bunker, 154.

cities, QL19 passed through An Khe, From Pleiku it continued into Cambodia. The trip from An Khe to Qui Nhon was approximately 64 kilometers (about 40 miles[67]).

Eleven kilometers east of Camp Radcliff the highway ran through a steep mountain corridor known as An Khe Pass. Here the treacherous motorway weaved its way amidst lofty mountain peaks. In some spots, sheer slopes rose up on both sides of the pavement. In other places, precipitous bluffs dropped off on one side. In other locations, the slopes mimicked gentle rolling hills. In order to avert ambush in the notorious passage, the military habitually defoliated the hills on either side of the pass by spraying them with the notorious herbicide known as Agent Orange.[68] The name Agent Orange was derived from the orange stripe on the herbicide's storage cans.[69] Defoliation left copious quantities of dead gray tree trunks, their vertical shafts standing watch over green grass and low bushes.

The 2½-ton M35 truck in which I rode had been field modified in Vietnam. The transportation unit had removed the canvas cover so that the truck bed was open to the sky. This allowed commuters to observe the surrounding terrain for enemy ambushes. Steel PSP planking was affixed to the truck bed's sides in order to increase the side's height. The floor of the bed was covered with a layer of sandbags. As we rode through the infamous pass, nervous eyes combed the landscape for signs of the Viet Cong.

The vehicle in front of my truck was a gun truck. These vehicles

---

67. 1 kilometer is equal 0.62 miles. Hereinafter most distances are given in kilometers since that measurement was used by the army in Vietnam.
68. Kelley, *Where We Were in Vietnam*, 5-11.
69. *Encyclopedia of the Vietnam War*, 2000 ed., s.v. "Defoliation" by Charles J. Gaspar, 93; United States Department of Veterans Affairs website (https://www.publichealth.va.gov/exposures/agentorange/). Accessed March 12, 2018.

were field improvised for convoy escort. Although the army did not authorize these combat vehicles, every level of command in Vietnam accepted them as essential for convoy security.[70] Armor sides were fabricated on the truck bed. This protective covering was diverse. Some gun trucks had steel plates welded together to form their sides. A few positioned a M113A1 APC (Armored Personnel Carrier) body onto the bed. The primary armament on these vehicles was M-2 .50 caliber machineguns. Some also added M-60 machine guns or a XM134 7.62mm minigun.[71] These trucks were painted black or olive drab. They were given distinctive names by their crew. Appellations like "Canned Heat, "Eve of Destruction," "King Kong," and "Pandemonium" were derived from pop culture. The name was painted boldly on the armor side in much the same fashion and style as nose art adorned airplanes in World War II.

Before riding in this convoy, I had limited opportunity to observe the civilian population of Vietnam. As the convoy churned down QL 19, we grappled with significant civilian traffic. Traffic laws seemed nonexistent. Automobiles occasionally slipped in between the convoy's vehicles, weaving in and out of line as they sped to get around us. Cars, buses, and trucks also passed going in the opposite direction. As they approached, they too were seen weaving in and out, using both lanes in a scary game of chicken. These vehicles generally appeared old and shabby. Their colors tended to be faded pastels. Various bags

---

70. Major Dean Dominique, "Gun Trucks: A Vietnam Innovation Returns," *Army Logistician* (January-February 2006) [http://www.almc.army.mil/ALOG/issues/JanFeb06/gun_trucks.html (website)] Accessed June 29, 2017.

71. United States Army Transportation Museum website (http://www.transportation.army.mil/museum/transportation%20museum/harden.htm). Accessed June 29, 2017.

and luggage of every conceivable form were stacked high and tied down with ropes on top of bus roofs.

Sporadically the convoy moved past pedestrians going both directions alongside the highway. I was taken aback when I saw an old woman squatting along the side of the road answering nature's call. She was wearing the ubiquitous black silk Áo bà ba outfit and a conical straw hat known as a Nón lá. The Áo bà ba is the traditional Vietnamese "pajama" attire. It consists of pants and a long shirt. Below the waist, the shirt is split up each side. My shock was symptomatic of the culture clash between Americans and the Vietnamese. The differences between the two traditions frequently led to misunderstanding and distrust. A feeling of cultural superiority was common among American soldiers serving in Southeast Asia. Derogatory racial labels such as "gook" or "dink" were employed almost universally by Americans. Because I possessed little interaction with the South Vietnamese population, I cannot accurately comment on their attitudes towards their American guests. However, I must assume some resentment surely developed in response to American arrogance. Clearly, misunderstanding between the two peoples hampered the war effort.

The landscape changed dramatically as we got closer to Qui Nhon. The rugged hills of the Central Highlands gave way to the flat open terrain of the coastal plain. Occasional palm trees replaced thick hardwood jungle. The territory became more populated. The highway passed through several small villages. People could be seen standing in the shallow rice paddies laboring to grow crops. The incongruous sight of a small boy driving a massive water buffalo with a bamboo pole is etched in my memory. The behemoth beast could have squashed the tiny child with one step. Instead, it responded obediently to the child's swats and pokes.

I remained in Qui Nhon for only a brief time. During the stay, I read a paperback western novel and built a plastic model kit. Once the predictable army paperwork was filled out in triplicate, I moved on to my next assignment. On 23 October 1970, after slightly over one month, I was relieved from duty with the Fourth Infantry Division and assigned to the legendary One Hundred and First Airborne Division. My brief tenure with the "Red Warriors" of the "Ivy Leaf" had been extremely peaceful. Nevertheless, the men in my platoon had taught me well about life in the bush. "Bush" was slang for the jungle or any area outside a base camp or FSB; it also was called "the boonies." The insight of these veteran grunts was not theoretical. They had learned it through experience. Passed on to me, their skills better prepared me for combat. Alpha Company's commanding officer, Captain John T. Plight, rated my conduct and efficiency both as "Excellent" in my clearance records. However, I had not yet really been out in the bush. For the infantryman, the bush designated his primary residence...and his combat arena.

I traveled from Qui Nhon to Da Nang in a C-130 Hercules cargo plane. This remarkable prop aircraft entered service in 1956 and remains in service 64 years later. Four Allison turbine engines powered the C-130E used in Vietnam.[72] The C-130 was built by Lockheed Georgia in Marietta. My father and my future father-in-law both worked on their production. I could not help a moment of contemplation. Had either worked on this particular C-130? Could they have imagined I would one day fly inside it over a war zone?

I flew military standby from Da Nang to Phu Bai. Consequently, I enjoyed a two-day layover in Da Nang. I spent all of it in the airport, but that was better than being in the boonies. During that time, a stunning blonde-haired woman passed through the terminal. She was ushered quickly through the waiting area by her entourage. She

---

72. U. S. Airforce website — http://www.af.mil/About-Us/Fact-Sheets/Display/Article/104517/c-130-hercules/. Accessed June 29, 2017.

paused in front of me only briefly for a pedestrian traffic jam. She glanced in my direction and I fancied that she smiled at me. I have often wondered since then who was she? I frequently speculated that she might have been Pam Eldred, the 1970 Miss America. However, Miss Eldred's visit to Vietnam ended a month or so earlier. Or perhaps she was actress Chris Noel, the "Voice of Vietnam." Chris Noel had her own radio program on AFVN (American Forces Vietnam Network[73]) at the time. She would leave Vietnam later that winter after more than four years of entertaining American troops. However, the young woman likely was a less known USO performer. Between 1965 and 1972, the USO (United Services Organization) organized 569 shows and 6,500 performances in Vietnam or at military hospitals in the Pacific.[74] The young woman stood out because of her beauty and her European heritage. In the crude slang of the troops, she was a "round eye," in contrast to an Asian "slant eye."

The young woman in a very short civilian skirt and white go-go boots[75] provided a welcome distraction from a troubling thought. I had been assigned to the 101st Airborne Division. I wore the Screaming Eagle patch on my left sleeve. Therein lay the genesis of my trepidation. The division had a reputation... a reputation for being where the fighting was hottest. Moreover, the division operated in Military Region 1 (more commonly referred to by its ARVN[76] designation

---

73. Touchstone Pictures' 1987 movie *Good Morning, Vietnam*, starring Robin Williams as an AFRS (Armed Forces Radio Service) DJ is based upon the AFVN radio operation.
74. USO website (http://www.uso.org/uso-entertainment-history.aspx). Accessed June 29, 2017. Accessed June 29, 2017.
75. Go-go boots were a knee high ladies boot with a square toe and low block heel that were very fashionable in the late 1960s and early 1970s.
76. ARVN [pronounced *R-vun*] was an acronym for Army of the Republic of Viet

"I Corps"—pronounced *eye core*). I Corps also had a reputation. It was considered the most blood-stained region in Vietnam (45% of all American deaths in the Vietnam War occurred in I Corps). Waiting in an airport—with nothing to do—allowed me too much time to think about my future. Under those circumstances, a person's thinking turns melancholy. The odds for a grunt to escape Vietnam without death or injury were not good. Serving in I Corps with the 101st made the odds even worse. I remembered my basic training company commander telling us we all were going to die in Vietnam. Accordingly, erotic contemplations about an unidentified American girl produced better mental health than did reflections on my potential for lying in a flag draped coffin. So I tried to divert my thoughts from any morbid predictions about my fate and fill my head with reminiscences of my fleeting encounter with that attractive young woman.

Inevitably, after over 30 hours, the delay in Da Nang ended and I was on another C-130. This time the destination was Camp Eagle at Phu Bai. I reported to headquarters of the 101st Airborne Division at Camp Eagle on November 5, 1970, as per orders.[77] I spent a brief stopover in the 101st reception station before being sent to Camp Evans. Evans was located 24 kilometers northwest of Hue on Route 601 near the small hamlet of Phong Dien. Travel once again was in a truck convoy. To my amazement, the trucks in this convoy had not been hardened or modified and no gun truck escorted them. Although I was much farther north, the danger seemed less. The trip seemed more like a tourist excursion than a combat movement. I have often thought since that day about the disparities in the two convoys. Perhaps the placidity of the latter reflected the effectiveness of the 101st Airborne in pacifying

---

Nam, the South Vietnamese army.

77. Headquarters 4th Infantry Division, Special Orders Number 296, Extract 221. TC 200, 23 October 1970.

the area it controlled. More likely it was circumstantial. The terrain was far less favorable to ambush than An Khe Pass. Moreover, the distance that the convoy covered was much shorter, about 20 miles on a major highway. Furthermore, much of it traversed heavily populated areas. Therefore, security of the roadway during daylight was more feasible.

The route passed through the Imperial Capitol of Hue. Three sites in that city were particularly memorable. First was a Catholic Cathedral. Its massive white facade with its enormous bell-shaped stained glass window dominated one intersection. The architecture with its interesting mixture of gothic form and oriental styles was striking. It looked out of place in the ancient Vietnamese city, but it certainly did not belong in Europe either. Second was the old imperial Citadel with its moot and blackened brick walls. Third was the bridge over the Perfume River. It was memorable because of its sheer size. Visible reminders of the 1968 Tet offensive remained throughout Hue. Scars from bullets pot marked numerous buildings. A few burnt out vehicles still littered right of ways. My trip finished without incident. I arrived at Camp Evans and the Screaming Eagle Replacement Training Section (SERTS) for orientation and training before being assigned to an infantry company with the elite 101st Airborne Division (Airmobile).

3. RAKKASANS AND MONSOON

Assignment to the 101st Airborne Division

SCREAMING EAGLES

By United States Army standards, the 101st Airborne w is a relative new comer with a history that only dates back to 1942. Despite its brief existence, the division has earned a fabled legacy in military history. In World War II, the 101st Airborne Division jumped into Normandy the night before D-Day. In September 1944, it jumped into Holland as part of Operation Market Garden. In December of that same year, the division was sent into Belgium to plug the gap created by the German offensive that came to be known as the Battle of the Bulge. During that battle, the 101st was surrounded and cut off in the town of Bastogne. On December 22, 1944, the German army called for the division's surrender and threatened its total annihilation if the Americans refused. General A. C. McAuliffe, acting division commander, responded with a single word, "Nuts!"[78] McAuliffe's answer

---

78. Leonard Rapport and Arthur Northwood, Jr., *Rendezvous with Destiny: A*

epitomized the unit's esprit de corps and sealed its reputation as an elite combat force.

Twenty-five years later, the 101st Airborne Division (Airmobile) was fighting in another war on the opposite side of the globe. Instead of being paratroopers, the grunts in the division now were experts in airmobile operations.[79] The bitter cold snow of Bastogne had been replaced by the sweltering jungles of Vietnam. Silk parachutes had been replaced by Huey helicopters. Instead of all volunteers, the ranks now included draftees. Nevertheless, the impervious esprit de corps and elite status endured. The Screaming Eagle insignia or the word "NUTS" was painted on every blank wall at the replacement center and at SERTS. The heritage and supposed superiority of the division's soldiers were drilled into every replacement.

One manifestation of this pride was the 101st Airborne Division (Airmobile) patch worn on the left shoulder of each man's uniform. As Ambrose observed, since Bastogne soldiers in the division have worn it "with the greatest of pride."[80] By 1970, the other divisions of the United States Army used a subdued green and black patch on their jungle fatigues. Not so the 101st Airborne. It retained the colored patch used in World War II. The division patch was a black

---

History of the 101st Airborne Division (n.p.: 101st Airborne Division Association, 1948), 509-14.

79. Today the *airmobile* designation has been superseded by the terminology *air assault*. The 1st Air Cavalry Division was the United States Army's first airmobile division. The 101st Airborne was the second airmobile division. The division changed from a parachute jump qualitied unit to airmobile during the Vietnam War when the effectiveness of air assault by helicopter had proven its value in combat operations. Today the 101st is the only air assault division in the world.

80. Stephen E. Ambrose, *Band of Brothers* (New York: Simon & Schuster, 2001), 219.

shield with a white eagle's head. A red tongue extended out of the eagle's open yellow beak. A black and yellow "Airborne" tab arched over the top of the shield. Hence, the official sobriquet of the division was the "Screaming Eagles." However, in Vietnam the patch sometimes derisively was mocked as the "puking buzzard." The irreverent moniker only added to a grunt's pride in belonging to the division. The North Vietnamese confused the eagle with the more familiar chicken and referred to the 101st Airborne as the "chicken men." The appellation was not derisive. NVA and VC units were often warned to avoid combat with the "chicken men" for fear of the adverse outcome.

In his book *Into Laos: The Story of Dewey Canyon II/Lam Son 719*, Keith Nolan maintained that the division's capability in early 1971 was "arguably the best combat unit still in Vietnam."[81] The grunts of the 101st thought they indisputably were the best combat soldiers in the world. Nolan attributed the division's combat effectiveness first to the high quality of its commanders. Second, the division operated in an area where contact with the enemy was made on a regular basis. Survival demanded the grunts maintain combat efficiency. Finally, the division could see positive results from their war effort. Enemy forces had been progressively pushed further and further away from the division's base camps and the civilian population in the region. Therefore, the grunts believed they would win every fight in which they were engaged.[82]

Although cynical disillusionment with the war was clearly present in the 101st Airborne by 1970, it did not yet influence unit cohesiveness. In reality, much of the esprit de corps came from the bonds forged at a platoon level. A platoon theoretically consisted of

---

81. Keith Nolan, *Into Laos: The Story of Dewey Canyon II/Lam Son 719* (Novato: Presidio Press, 1986), p. 39.

82. *Ibid.*, p. 39.

approximately 40 men. In reality it averaged closer to 30 most of the time. A grunt's platoon was his family. He lived with these men 24 hours a day, seven days a week. They knew each other thoroughly. They depended upon each other for survival. They shared everything they had. A grunt's personal identity—his race, his education, his economic status, his home state, his civilian occupation—diminished and his identity with the platoon amplified as they devoted time together. One's own life was secondary to the lives of one's buddies within the platoon. And with few exceptions, every grunt in the platoon was his buddy. Soldiers took pride in belonging to the fraternity of the men with whom they lived on a daily basis.[83] This small unit pride intrinsically is coupled to a platoon's identification with its higher organization, specifically its company, battalion, and division. In the presence of soldiers from other outfits, the Screaming Eagle's grunts frequently displayed a cocky arrogance forged by the unit's brief, but illustrious history. This *Esprit de Corps* created a reputation that mystified many outside the 101st Airborne. A soldier from another infantry division once told me, "You guys are crazy." We took full advantage of the division's aggressive reputation to improve our lives whenever the opportunity presented itself.

Later in my tour, I returned home on a two-week leave. While waiting for transportation stateside, two other enlisted men from the 101st and I went into the Air Force Officers Mess at Tan Son Nhut

---

83. The strength of the bond created within a platoon became very apparent to me when I met Russ Kagy and Alan Davies in 2018 for the first time after 47 years. They both were in my platoon, but neither were my closest buddy. Although we had each pursued different paths in civilian life, and only made contact with each other again recently, we related to each other as if four and half decades were only four and half hours. The 47 years apart could not devalue the regard we held for each other. Later communication with Kevin Owens was equally warm.

Airbase. The Air Force NCO at the door observed the absence of any rank insignia on our uniforms and refused to allow us inside. When we acknowledged smugly that we were enlisted men, he was insistent that we leave.

"Only officers are fed in this facility!" the NCO said emphatically.

However, we too could be insistent. We stubbornly refused to go. Then he took note of our Screaming Eagle division patch on the left shoulder and the blue metal CIB (Combat Infantry Badge) on our chests. The CIB was awarded only to soldiers with an infantry MOS and who had engaged in active ground combat. Vietnam infantrymen generally viewed it as the second most prestigious medal awarded by the army. Only the Medal of Honor held more general prestige (but many of its recipients were dead when it was awarded). The CIB signified its wearer belonged to an extraordinary fraternity, those who had fought the enemy in the bush. Wearing the silver and blue dress badge on jungle fatigues was unauthorized, but it added to the intimidation of our trio.

In the end the airman backed down, saying, "If you promise not to destroy the place, I'll make an exception for you."

Our bluff had paid off. He seated us in one corner of the room and I enjoyed one of the best meals I ate in Vietnam. After dinner we left quietly and returned to our quarters for a good night's sleep.

THE SCREAMING EAGLE REPLACEMENT TRAINING SECTION

All new replacements to the 101st Airborne Division were required to take the division training course at SERTS (Screaming Eagle Replacement Training Section). SERTS was located at Camp Evans. The course normally was two weeks long. Because I was an in-county transfer, my time at SERTS was abbreviated. Although I had not yet been in the bush, due to the 4th Infantry patch on my right sleeve I was viewed as a veteran, not a cherry or FNG. Primarily the portion of the curriculum I received evolved around airmobile training and division procedure. At division orientation, the briefing officer explicitly stated many of us would be killed or wounded. Despite Vietnamization, the division remained aggressive in its combat operations. It operated in one of the most hostile regions in Vietnam, the mountain ranges and valleys along the Laotian border. The North Vietnamese ambition to take Hue only could be achieved if their forces first destroyed allied opposition in this region. The 101st Airborne Division comprised the leading ground element of that opposition.

I had been exposed to rappelling in AIT. At SERTS I got to do it. Rappelling down a 40 feet high wall and off the skids of a helicopter were unforgettable experiences. The rappeler wore a rope harness. It was affixed to a rappelling rope with a medal "D" ring. The rappelling rope ran between a person's legs to the ground. One hand grabbed the rope below the buttocks. The other hand held it in front of the chest. The speed of descent was controlled by applying greater or lesser tension on the rope below the "D" ring. Sufficient force produced a complete stop. Nylon ropes rendered timing the stop crucial. Stopping the descent too soon resulted in a yoyo effect. Instead on planting one's feet on the ground, the unfortunate jumper bobbed up and down a foot or so above the ground. Braking too late ended with a severe jolt as the legs banged into the ground. The procedure was simple. The soldier stood on the edge of the tower or the skid of a Huey facing away from the direction of his fall. Then he leaned back and gravity took over. Down he went. On the tower he hit the wall with his feet and then pushed his body away from the wall with his legs as he regulated his descent.

Rain was falling the day I rappelled from the tower and the

wall was slick. I slipped on the wet wood and ended up suspended upside down 30 feet above the ground. For a moment I envisioned my body splashed all over the ground below me. Somehow—I'm not sure how—I righted myself and completed my descent without breaking my neck. I completed the jump from a helicopter flawlessly.

Classes in 101st Airborne Division SOP (Standard Operating Procedures) and courses on NVA SOP were taken. I don't remember much else about SERTS, but five days after arriving, my orientation was complete. Thereafter I was assigned to Charlie Company, 3/187 (enunciated as third of the one-eight-seven), officially Company C, 3rd Battalion of the 187th Infantry Regiment (Airmobile). As a novice the distinct pedigree of this battalion was unknown to me. As far as I was concerned at the time it was just another unit designation.

RAKKASANS

The 187th Glider Infantry Regiment had not been part of the 101st Airborne Division in World War II. Instead it served with the 11th Airborne Division in the Pacific Theater. It had fought in New Guinea and the Philippines. During its occupation of Okinawa in 1945, the 187th became the first airborne unit to jump on Japanese soil. There the Japanese gave the unit its appellation "Rakkasans." Rakkasan roughly translates "falling down umbrella," a vivid description for a parachute. During the Korean Conflict, the regiment operated as an independent unit, the 187th Airborne Regimental Combat Team. It made two major combat jumps in Korea. Hence its Pacific Theater and Korean War heritage was unique within the 101st Airborne Division. In Vietnam, the 3rd Battalion of the 187th Infantry became the most decorated airborne battalion of the war.[84] It remains one of the most decorated battalions in United States Army history.[85] The regiment's colors proudly fly 15 Citations for Valorous and Meritorious service and 23 Battle Campaign Streamers. No other airborne unit can equal that record.

Just 17 months before I was assigned to it, Charlie Company earned the distinction of being the first American unit to occupy the crest of Dong Ap Bia, commonly called Hamburger Hill.[86] The company suffered around 80% casualties in that bloody battle and the legacy of the fight still lingered heavily over Charlie Company in the fall of 1970. It secretly was whispered that the mission of the battalion in 1970 was not to get anyone killed. The army did not want any more of the negative publicity comparable to the news reports that came in the aftermath of the controversial battle.

The 3rd Battalion of the 187th Infantry was based at Camp Evans, not far from SERTS. Therefore, reporting to my new unit consisted of a short ride in the back of a deuce and half truck. The battalion area was a compound of neat, painted wood structures. The lower half of most buildings was painted forest green. The upper half was white. All of the buildings had tin roofs decorated with scattered sandbags. Unlike Camp Radcliff, however, the buildings did not have dirt-filled 55-gallon drums parallel to the outside walls.

---

84. The website of the Rakkasan Association (http://www.rakkasan.net/index.html) and military unit history (http://www.military.com/HomePage/UnitPageHistory/1,13506,100793%7C778079,00.html). Accessed February 8, 2016.

85. Frank Boccia, *The Crouching Beast: A United States Army Lieutenant's Account of the Battle for Hamburger Hill, May 1969* (Jefferson: McFarland & Company, Inc., Publishers, 2003), p. 4.

86. Samuel Zaffiri, *Hamburger Hill: The Brutal Battle for Dong Ap Bia, May 11-20, 1969* (New York, New York: Ballantine Books, 1988), p. 258.

The centerpiece of the battalion area was a small painted statute of a bald eagle standing on a globe of the world. It was flanked by two flagpoles. The United States flag flew from the pole on the left and the Republic of Vietnam on the right. Behind the two flags was a large red Japanese *torii*, the informal symbol of the 187th Infantry. The *torii* was the gateway to a Japanese Shinto temple. It has two posts and two crosspieces. Its usage by the 187th symbolized the unit as a gateway of honor. The upper crosspiece had "RAKKASANS" painted in white uppercase letters. "3rd of 187th" was painted on the lower crosspiece. The regimental crest hung on the wall of the building directly behind this display.

Company orderly rooms lined one side of the battalion parade ground. A PSP sidewalk connected the orderly rooms. Outside each structure, hanging on a white signpost, was a wooden blue infantry guidon designating each company, A through E. The orderly room was the company administrative office. The company's commanding officer, first sergeant, and clerk had desks in this building. I reported to Charlie Company's orderly room and the first sergeant assigned me to first platoon.

After completing the requisite paperwork, I wandered down to the platoon's hooch. Unlike the orderly room, these flimsy buildings were painted solid green. The wood-frame structure utilized 2x4 studs at 2 feet on center. 4x8 sheets of plywood were nailed horizontally to the bottom half of the wall. Inside, this portion of the wall was paneled with unpainted vertical boards scavenged from wooden crates. A wire screen net covered the top half of the walls. Nails in the 2x4 top plate of the wall framing provided impromptu pegs on which to hang uniforms or field equipment. A door was located in the center of both end walls. The floor was plywood on wood joists. Exposed rafters supported the ever-present tin roof. A couple of bare light bulbs were suspended from the rafters to provide limited lighting at night. A few electrical outlets were spotted on the side walls. An aisle ran down the center of the room with a row of army cots on each side.

During this period I suffered from the delusion that my tour would be shortened. The war was ending and troop strength in country was dropping as units were redeployed stateside. Army gossip predicted DEROS dates would be cut by two months as part of the American withdrawal from Vietnam. I foolishly listened to the rumors and wrote letters that I would be home for my twenty-first birthday in 1971.[87]

COMBAT OPERATIONS IN THE MONSOON SEASON

In June 1968 General Creighton Abrams, replaced General William Childs Westmoreland as Commander, MACV (Military Assistance Command, Vietnam). The new American commander changed tactics, replacing multi-battalion sweeps with small-unit patrols in an effort to block enemy access to the civilian population.[88] Consequently Charlie Company commonly operated alone in the field. Frequently its platoons were dispersed widely in its AO (an acronym for Area of

---

87. *E.g.*, in a letter to my parents dated "14 Dec 70," I wrote, "It really looks like my Deros date may change from 13 Sep 71 to Jul 71…That means only seven months left. WOW!"

88. *The Encyclopedia of the Vietnam War: A Political, Social, and Military History*, ed. Spencer C. Tucker, 2000 ed., s.v. "Abrams, Creighton" by Kenneth R. Stevens, pp. 1-2. Reduction of the number of combat units in Vietnam and the Vietnamization policy were factors in the change [Stanley Karnow, *Vietnam: A History* (New York, New York: Penguin Books, 1997), p. 610].

Operations). The term designated the geographic locale in which a unit carried out its assignment.

During my tour, Charlie Company operated almost exclusively in "Free-Fire Zones." Free-Fire Zones were designated areas in which the civilian population had been relocated elsewhere in Vietnam. This meant anyone the company encountered in the field was considered hostile and therefore should be engaged. Yet contact with the enemy during this monsoon season usually was on a small scale. This was in part due to the success of the 101st Airborne Division's combat operations in the region. NVA units had either been destroyed or pushed back into the A Shau Valley or Laos. The combat effectiveness of the division to date made the use of significant sized units not viable for the enemy at that time. Consequently they rarely operated in anything larger than a squad or a platoon, frequently moving only in fire teams. Their primary mission for the moment seemed to be harassing American units and rebuilding their strength for subsequent offensive operations after the American withdrawal was complete.

Engagements with these small units tended to be more accidental than intentional. An instance from mid-November 1970 illustrates the haphazard nature of such engagements. After spending two nights on one hilltop, Charlie Company moved down the summit to a new PZ for extraction. PZ was an acronym for Pickup Zone, a piece of unobstructed ground large enough for a helicopter to land and pickup troops. A Landing Zone's acronym was LZ. A LZ differed from a PZ in that in an LZ troops debarked out of the helicopter, whereas in a PZ the troops loaded onto the helicopter. In Vietnam these acronyms were used almost exclusively rather than the word designation. Enlisted men in Charlie Company rarely distinguished between a LZ and a PZ, routinely calling both an LZ. The move on this November day was necessitated by operational requirements for an LZ (*i.e.*, PZ). LZs required an open area accessible to the UH-1D/H Huey helicopters that carried supplies and transported troops. The landing zone itself obviously had to accommodate the rotary span of at least one Huey. For extraction of ground forces, multiple helicopters were preferred. Otherwise the risk to the last loads out was greatly compounded. Likewise, the approach and takeoff needed to be clear of obstruction. Vertical landing and takeoff usually was impractical.

Although an LZ on the crest of the ridgeline would have been ideal, the dense triple-canopy jungle there required too much effort to cut one without engineers and their equipment. A large open expanse was located down the slope and an LZ was set up there. However a bellicose rain storm grounded the helicopters, preventing the extraction of the company. Consequently the company was forced to return to its previous night's Night Defensive Position, or NDP. NDPs commonly consisted of a series of foxholes laid out in a circle. This arrangement enabled the grunts to lay down defensive fire for 360 degrees from the NDP perimeter. The CP (Command Post, the unit commander's position) was set up in the center of the circle. Contingent upon terrain and circumstances, a shallow sleeping hole also might be dug behind each foxhole. Claymore mines, trip flares, and mechanical ambushes were set out beyond the perimeter in strategic sites.

On this occasion, returning to an old NDP was determined to be the preferable option from the few available poor alternatives. Being only a private, I was not privy to the rationale behind the decision. The absence of a suitable location for a new NDP near the proposed LZ likely provided the primary justification for the decision. Furthermore, the company lacked adequate daylight to dig new foxholes. In the brief time since the company abandoned the NDP, it seemed unlikely that the enemy realistically would have occupied the site or set up booby-traps in the abandoned holes. Returning to the old site violated the unit's standard operating procedures. So our return was aberrant and therefore totally unexpected by the enemy.

Second platoon led the way back up the hill. First platoon trudged up the trail behind them. Third platoon fell in at the rear of the column. The rain and heavy foot traffic had transformed the trail into gooey mud. It sucked one's boots down into the slippery slime, making movement laborious and slow. I was approximately a third of the way up the slope when I heard firing from the front of the column. Both M-16 and AK-47 fire was clearly audible. Instantly, first platoon's pace accelerated to double-time as we raced toward the firing. The weight of our heavy rucksacks and the struggle with the primordial slime worked against the adrenaline spawned by the gunfire. With labored breathing and cramping muscles we approached second platoon's position near the old campsite. We yanked the quick release strap on our right shoulder and dropped our rucksacks; quickly first platoon deployed into a firing line adjacent to second platoon. Before we could join the firing, an order to ceasefire was given. The whole episode had lasted less than five minutes.

As we moved on up the hill to the NDP, we passed the site of the shooting. A single corpse wearing black pajamas lay on the ground in front of us. He stared blankly up toward the sky. Blood saturated his abdomen and flies already buzzed around the crimson gore. He wore a pair of Hô Chí Minh sandals cut from an automobile tire. He had no headgear or accouterments. He was stockier than most Vietnamese. Recovered papers on his body alleged he was a Chinese medic. A few soldiers in second platoon stood nearby. One inspected the dead man's AK-47. An extra magazine was fastened to the magazine in the weapon with black electrical tape. To reload the weapon its shooter only needed to pull the empty magazine from the chamber, flip it over, and insert the full magazine into the vacated chamber.

In Vietnam, a FNG or cherry, the newest guy in the unit, commonly performed most unpleasant tasks. Russ Kagy and Alan Davies were both from California and were new arrivals in first platoon. Alan was a lanky kid from San Diego. The guys in the platoon soon labeled him as Uncle Al, after the host of a popular children's television show. Russ was stockier than Alan. He came from Los Angeles. Russ had bushy black hair and a thick mustache. The CO assigned them the chore of burying the enemy corpse. They pulled out their entrenching tools, unfolded the shovel blade, and locked it in place by twisting the nut tight. They hastily dug an insubstantial hole and rolled the body in it. The two disgruntled gravediggers tossed just enough dirt over the dead body to cover it. However, one forearm and hand inadvertently protruded straight up out of the shallow grave. A few grunts chuckled at the bizarre spectacle. A few others objected to the macabre humor. Caught between the opposing viewpoints of the company's veterans, the baffled cherry kicked the arm in an impulsive effort to force the limb under the ground quickly since the platoon was already moving out. His kick flattened the offending appendage but broke the corpse's bone in the process.[89] The spontaneous stroke ended the discord in an audible chuckle. Alan and Russ tossed a thin layer of dirt over its broken arm and the platoon moved back into its previous night's foxholes. Russ's decisive reaction cemented his place in first platoon's lore. He had taken a giant stride in moving from "just another cherry" to a beloved family member.

This was the first dead body I encountered in Vietnam. I was to see many more in the months ahead. The dead included both American and Vietnamese. In time I would grow numb to seeing corpses. They would become a part of my normal reality. Part of the tragedy in ground combat is how callous soldiers can become to death. I did not have time to scrutinize this cadaver long. Nevertheless, the sight of the enemy corpse cast a somber shadow over me. This was not like

---

89. Russ Kagy, Alan Davies, and myself discussed this incident in detail at the 101st Airborne Division Snowbird Reunion in Tampa, Florida, February 8-10, 2018.

watching combat in war movies. This was real and it was dangerous. People were trying to kill you and you were trying to kill them One moment you were alive; the next you could be dead. It happened just that fast.

When the company abandoned the site earlier, three Viet Cong soldiers moved in looking for food. The Viet Cong, commonly dubbed VC, were guerrilla insurgents in South Vietnam. They typically dressed in the áo bà ba of the civilian population, a two piece silk pants ensemble. Whereas the civilians sometimes wore various colored tops with black pants, the VC preferred all black for stealth and subterfuge. Americans referred to this traditional Vietnamese attire as "black pajamas." In the area where the 101st Airborne Division operated, the VC had been decimated by 1970. Hence contact with VC was uncommon. Most hostile action involved the NVA (North Vietnamese Army[90]). The NVA was a well-disciplined and aggressive foe. Its soldiers primarily dressed in khaki or olive green uniforms and usually were well equipped. The presence of a Chinese advisor suggests these three VC were NVA replacements for a depleted VC unit in the populated coastal region.

Second platoon's uncharacteristic return to the NDP surprised the three guerrillas. Members of second platoon claimed they hit all three, and insisted two had been "shot up real bad."[91] However, only one enemy body was left on the hilltop and so the official body count was reported as one. Nevertheless the other grunts on the ground accepted the second platoon's version, two VC killed (one of the three had to have survived in order to carry off the second body). One

---

90. The official North Vietnamese designation of its army was "People's Army of Vietnam." Hence some recent works refer to the NVA as PAVN. I have used NVA because it was the terminology used exclusively in Charlie Company in 1970-1971.

91. Letter to my parents dated "3 Dec 70."

lieutenant in second platoon injured his finger slightly. There were no other American casualties.

The company commander called in two Cobras to shoot up the jungle where the VC had fled. This was my first occasion to observe Cobra gunships in actual combat.[92] The Bell AH-1G Huey Cobra was designed as a tactical aerial weapons platform. Its narrow frontal silhouette was only 38 inches at its widest point. The sleek fuselage enabled it to achieve speeds over 160 miles per hour in level flight. Tandem seating placed the gunner in front, giving him virtually an unlimited view. The pilot sat directly behind and above the gunner.[93] The pair of gunships that responded to Charlie Company carried rockets, a 7.62mm minigun, and an automatic 40mm grenade launcher. The minigun fired 6,000 rounds per minute.[94] Watching the two Cobra's firepower made a definite impression on this novice grunt. I wrote home, *"It sure makes you feel better to have them on our side."*[95]

In November, December, and early January, the weather became a grunt's biggest enemy. The weather in I Corps was dictated by the northeast monsoon season. The heaviest rainfall occurred between late October and early January. Unlike the hot monsoon rain in the southern regions of South Vietnam, the monsoon in the area where the 101st Airborne Division operated in 1970 was a cool rain.[96]

---

92. A Cobra participated in the mad minute demonstration during AIT. In that demonstration the army's entire arsenal was firing. Therefore, rounds from the Cobra blended with the firepower of other weapons and left no individual impression on me.

93. Simon Dunstan, *Vietnam Choppers: Helicopter in Battle, 1950-75* (Oxford: Osprey Publishing Ltd., 2003), pp. 111-13.

94. Stanton, *Vietnam Order of Battle*, p. 293.

95. Letter to my parents dated "3 Dec 70."

96. Asian travel website hosted by Samantha Brown and Stuart McDonald (http://

Grunts in the field stayed wet and miserable. Rain gear was of little value. It was restrictive and uncomfortable. Besides, the rubberized garments caused the wearer to sweat profusely. A grunt got as wet from perspiration wearing it as he did from the rain without it. Hence most grunts did not wear rain gear.

Today the verb "hump" can mean, "to carry something heavy with difficulty."[97] It possibly derived this sense out of the Vietnam experience. In the mid-twentieth century, the verb had two meanings. First, it meant, "to cause to assume the shape of a hump."[98] Second, it had the slang sense "to exert (oneself)."[99] Both were applicable to an infantryman in Vietnam. The heavy pack caused him to hunch forward, giving the appearance of a large camel-like hump on his back. The sheer weight of his equipment made movement difficult in the adverse tropical conditions of Vietnam. There the word acquired the sense of a military movement on foot. Units humping in the thick jungle vegetation generally kept in a single file formation with the grunts spaced approximately an arm's length or more apart. They were hot, sweaty, and exhausted. Their numb bodies mechanically stepped along like zombies.

Often the terrain over which Charlie Company humped in late 1970 was extremely steep hills and mountains. Consequently, short distances on a map might in reality entail significantly longer differences in actually. For example, a map might show a distance of one kilometer[100]. However, in the steep mountains where we operated the

---

www.travelfish.org/weather/vietnam). Accessed August 4, 2017.

97. Encarta Dictionary: English (North America) in Microsoft Word 2010.

98. *Webster's New Twentieth Century Dictionary*, 1979 ed., s.v. "hump."

99. *Ibid.*

100. Distances generally are noted in meters or kilometers. One meter is equal to 1.09 yards. One kilometer equals 0.62 miles or 3,280.84 feet.

difference in elevation made the actual distance walked much longer than on flat ground.

Humping in Vietnam jungles frequently required cutting a trail through the indigenous flora. Existing trails usually were avoided for fear of ambush and booby traps. The procedure for cutting trail generally was determined by the area's topography, the type of vegetation, and the perceived enemy threat. Whenever underbrush was thick, one or two men hacked out a path with machetes. The grunts rotated this draining task to balance out the fatigue of the unit. The unit could either move with the men cutting brush or send a small detail forward alone. In the former case, the column stayed on its feet, moving slowly with the cutters. In the latter case, the unit would sit while the cutting party slashed a trail for 20 to 25 meters. At that point the main force moved up to where the path had been completed. Then the unit repeated the process. This method generally was employed only in very thick underbrush. When cutting trails, the sound of machetes chopping through the undergrowth made secrecy of one's presence impossible. Therefore one grunt was detailed as security for the cutting party. He searched the foliage for signs of the enemy's presence.

Triple-canopy jungle allowed little sunlight to pierce through to its floor. Consequently little undergrowth hampered movement in some of these rainforests. So trail cutting was unnecessary in many jungles. Nevertheless, visibility rarely exceeded ten or twenty yards. Likewise many grasslands and cultivated regions did not need trails cut.

In environments where cutting trail was unnecessary, a two-man team led every movement. This team consisted of a point man and a slack man. The point man led the column. He was therefore the first American boots to walk into that space on a given day. He never knew what the deep jungle might hide in its dark undergrowth. If the NVA or VC lurked in its bowels, the point man was the grunt they spotted quickest. Intuitively the hidden enemy aimed their AK-47s at him first. For some grunts, the perils of walking point could be terrifying. For other grunts the adventure of walking point could be addictive. For most dependable point men, walking point was both. His responsibility was determined by the terrain and flora.

On trails and in open vegetation, the point man limited his survey to the few feet of ground directly in front of him. Here a point man's safety — and the safety of those who followed him — depended upon him seeing every inch of that ground. His primary responsibility was spotting land mines, trip wires, and various other booby traps. Shortly after joining first platoon, the platoon leader assigned me this role and I became a proficient point man. I quickly developed an inexplicable sense for spotting something wrong on the ground. A slight indentation in the surface, an unnatural placement of a stick or leaf, and other barely noticeable clues could signal danger. Even today — five decades later — I still walk with my vision glued to the ground where my next step is going. The point man's progress was slow and deliberate. Having determined the ground where his next step would be as safe, he methodically put his boot in the exact spot he had selected. Almost simultaneously he surveyed everything before him for any clues of the enemy's presence. Then he meticulously examined the ground directly in front of him. He searched for anything he might have missed previously and for new clues. Once he deemed the next few feet safe, he repeated the whole process.

The slack man walked directly behind the point man. His primary job was protecting the point man. His area of responsibility included the entire panorama in front of him. He looked for ambushes and other dangers to the side of and beyond the four or five feet of the ground in front of the point man. However, just as the point man repeatedly scanned the same area as the slack man, the slack man periodically eyeballed the path for signs the point man may have missed.

Hence the life and well-being of the point man and slack man intertwined. These two men functioned as a single entity. If they did not, the point team jeopardized the survival of their entire unit.

The last man in the column was the drag man. His responsibility was to guard against any threat from the unit's rear. He constantly turned around to survey the landscape through which the column had just moved. In the bush of Vietnam, "behind the front lines" did not exist. The ground behind an infantry unit's advance did not become secured territory. Rather the advance resembled a boat moving through water. The presence of the boat only displaces the water while the boat occupies the space. As the boat proceeds forward, the water reoccupies the vacated space behind the boat. Likewise, the only secured ground in the bush was the ground on which the infantry unit stood at that given moment. As an infantry column moved forward, the ground behind the column reverted into enemy hands. NVA or VC sometimes took advantage of the trails grunts created to mount sneak attacks against the rear of careless infantry units. Therefore the drag man's role was just as critical to the unit's effectiveness and survival as was the point team.

The monsoon compounded the difficulty of moving in the bush. At the time I described "moving" as "stumbling."[101] Wet garments added weight to one's load and chaffed one's skin. An olive drab towel was wrapped around a soldier's shoulders to pad the straps of his rucksack. The towel absorbed even more water, and hence more weight. The moist conditions likewise offered an ideal environment for ringworm and other infections. The monsoon rain turned the jungle floor into muddy sludge. The wet ground provided a slippery surface on which to walk. Slopes in the rugged mountain terrain in which Charlie Company operated were extraordinarily steep. Movement often

---

101. Letter to my parents dated "3 Dec 70."

meant periodically taking a step forward and sliding backwards one or more steps. Therefore the step had to be repeated.

The effort was like humping on a treadmill, always exerting an effort but making no progress. After two or three attempts, you grabbed a low branch or an extended hand and pulled your body forward. Every grunt experienced a fall at least once in his tour, usually more than once. When ascending a hill, these falls could knock down men behind the initial victim like pins in a bowling alley. Such incidents inevitably provoked an outburst of profanity from the unfortunate men who fell and chuckles from comrades who were still standing. Moreover, these tumbles added to the accordion effect of jungle movements. A column moved forward, stopped, and pressed together. The front of the column encountered improved conditions briefly and made rapid progress. Meanwhile the rear struggled, stretching the column out. By the time the rear reached the better ground, the point again grappled with adverse terrain and the column bunched back together as the rear raced to catch up.

Vietnam's abundant supply of snakes, leeches, insects, and countless other parasites added to the grunts' misery and danger. Military movements frequently entailed wading through rivers and streams, nicknamed "blue lines" after their representation on maps. These water courses were full of leeches that attached themselves to the skin. Grunts often buckled straps around their trouser legs just below the knee to prevent these bloodsuckers from getting to his crotch. A victim could not pull the leeches off. Trying to do so almost always left the head embedded in the skin. This in turn led to infections and other serious medical issues. Fortunately, simple removal procedures were readily available. Attached leeches could be detached by touching the vampire with a lit cigarette or squirting it with insect repellent. If the leech was not removed, it usually fell off after gorging on enough human blood to double its size. However, the creatures produce an anticoagulant that caused the bleeding to continue. The fungal infection ringworm (*dermatophytosis*) and bacterial infection boils were other common field conditions.

Insect bites were a constant, painful nuisance. However, mosquitoes posed a more lethal threat, malaria. Consequently, MACV required all American military personnel to take anti-malaria tablets. They swallowed a large orange colored tablet (*Chlorequine-prinaquine*) once a week. A white tablet (*Dapsone*) was consumed daily. Since malaria could develop after departure from Vietnam, troops continued an anti-malaria regime for two months after DEROS.[102]

Lieutenant Colonel Bryan J. Sutton commanded the 3rd Battalion of the 187th Infantry in late 1970. Like many commanders in the 101st Airborne Division, Sutton spent time in the field with his men. According to my experience, he treated enlisted men with respect and as individuals. Periodically he dropped in unannounced on a company or platoon in the bush. He observed rather than interfered with the units normal operations on those occasions. At the same time, he did not hesitate to address issues he perceived or his men brought up. One time during the monsoon, he asked about the large number of Claymore mines the platoon had used recently. Most of them had been broken open for the C-4 plastic explosive inside, but no one dared disclose this misappropriation of resources. The platoon had been having difficulty obtaining C-4 through regular resupply channels. Hence, we resorted to the Claymores because supply immediately filled orders for munitions. A resupply bird usually delivered such orders within a few hours.

C-4 is a white malleable explosive composition. Combat units routinely used it to blow down trees while cutting LZs or for destroying enemy bunkers. It came in rectangular blocks one and half inches

---

102. MACV Form 270-R (1 Jun 70).

by two inches and 11 inches long. Unless constricted in an enclosed space or ignited with a blasting cap, the plastic explosive could be burned safely as a fuel. It made an especially hot fire, ideal for brewing coffee or drying socks — both of which were important to a grunt's wellbeing in the field. In the absence of heat tabs, a small piece could cook C-rations.

Sutton interrogated the platoon but everyone maintained the Claymores had been blown at suspicious sounds and movements during the night. Despite his threats, no one deviated from the tale. I suspect the colonel knew how we used the Claymores but he didn't press the issue. He simply told us how much a Claymore mine cost and jested that using their C-4 for boiling water brewed an unduly expensive taste in coffee. Sutton issued his mild indirect reprimand with empathy for our plight and with an implication that he would correct the problem. Thereafter supply promptly filled our requisitions for dry socks or packs of C-4.

Sutton was a small man physically, but in the field he carried a standard rucksack. His load was lighter than the average grunt, but he carried everything he used himself. Although line companies of the 101st Airborne Division were still well disciplined combat units, the large number of draftees in the ranks endowed them with a certain anti-authority spirit. Colonel Sutton's unsolicited appearance in the field incited this insubordination to express itself in a futile effort to wear out the battalion commander by the day's hump. If a cliff blocked the route, we climbed it rather than go around it. If our path crossed a stream, we waded through it at the most difficult spot. The pace always was much faster than normal. However, Sutton was a former Green Beret and our ineffectual efforts only succeeded in making the day more grueling for ourselves.

Water is essential to life. The human body shuts down and death normally occurs within 2-3 days without proper hydration. Grunts commonly carried six or more quarts of water. During most of my tour, I carried one two-quart canteen inside my rucksack, three one-quart canteens attached to its outside, and one one-quart on my belt. A couple of times resupply airdropped rubber bladders of potable water to us. However, most of the time, we obtained drinking water from nearby streams or from rain runoff. Occasionally the water source contained foreign objects, such as mud, leaves, and insects. We removed these items by straining the water through the t-shirt on our back. Happily this was extremely rare. Drinking water from such unsanitary sources obviously was not healthy. Therefore the army issued iodine tablets to purify unhygienic water. However, the pills left a disgusting taste in one's mouth. Adding pre-sweetened Kool-Aid™ to the water disguised the unpleasant flavor. This precious flavored powder only could be acquired from family and friends in the United States. For that reason my letters home frequently included the petition "send Kool-Aid."

So Pamela and my parents routinely mailed me "care-packages" containing packs of Kool-Aid™ and other treats unavailable in Vietnam. The arrival of a care-package made you feel like a small child on Christmas morning. I unceremoniously ripped open the package in anticipation of discovering the unidentified goodies inside. I was particularly delighted when Mother or Pamela sent homemade cakes and cookies. These treats often were damaged in shipping, but broken cookies and smashed cakes still tasted great. Whatever bakery items anyone received were shared with the other grunts in his squad or platoon.

Monsoon conditions affected our NDP preparations. Whenever possible, we dug foxholes, but they usually filled up with water during the monsoon. The heavy rain made digging out sleeping positions impractical. Instead, grunts erected temporary hooches with ponchos. First, they tied one end of a bootstring around the head cover to seal the opening in the center of the poncho. Twelve inches or more of the other end of the bootstring hung free. Then the grunts stretched out the poncho about two feet above the ground, having secured the four corners to nearby bushes or stakes. The free length of shoestring connected to the head cover was tied to a supple branch directly above the poncho, thereby stretching the poncho into a pyramid shape. Typically, the occupants spread out another poncho on the ground beneath the cover with the outside down. This second poncho served as the hooch's floor. The provisional edifice provided cover for two men and all of their equipment.

A four-man position could be fabricated by overlapping two ponchos over a tent ridgepole. An appropriate size limb of a bush or tree functioned as the ridgepole. A four or five foot stick was cut and laid across the outside of the hooch perpendicular to the ridgepole. The poncho's head covers were tied shut and stretched to each end of this second stick. These hooches also could protect one from the hot tropical sun during the dry season.

Each soldier carried a gray air mattress. They inflated the mattress by blowing into the plastic stem at the top corner. Troops arranged the blown up mattress side by side under the poncho tent. In the steep mountains where Charlie Company operated, grunts made every effort to site the head uphill. They sometimes drove stakes into the ground at the foot of the air mattress. The mattress often slid downhill in the wet conditions otherwise. A poncho liner completed a grunt's sleeping kit. The poncho liner was a lightweight nylon quilt that replaced the standard olive drab wool blanket in tropical climates. These woodland camouflage pattern covers were a grunt's closest item to luxury in the field.

The security and safety of a unit in the bush necessitated guards be posted at night. Guard duty, therefore, was obligatory for all grunts in the field. Only the medic was exempt from this fatiguing duty. Guards typically rotated two-hour shifts. Under ordinary circumstances, one fourth of a unit always was awake on perimeter guard while it was dark. In areas inhabited by large numbers of NVA, the number of guards might increase to half the unit. Some conditions triggered a total alert in which everyone was awake, armed, and in their fighting position. Although this last level of security could not be sustained indefinitely, it might keep everyone awake overnight or longer. Additionally, SOP (standard operating procedure) compelled this highest alert level be maintained daily during the period just before and after dawn. First light appeared around 0500[103] hours (5:00 o'clock in the morning in civilian time). So at 0430 everyone was in their fighting position, staring blindly into the menacing darkness, fully equipped for combat with their weapon pointed at the latent enemy. By 0600 the sun was fully up and the daily routine got under way.[104] For this reason grunts customarily averaged six hours or less of sleep each night. Less than half of that slumber was uninterrupted sleep. Consequently, physical exhaustion was a grunt's perpetual companion.

Daylight security tended to be less formal whenever a unit remained in its previous night's NDP. A few soldiers always were on alert as sentinels. The darkness no longer hid possible clandestine enemy encroachments. Therefore the number of guards posted was reduced significantly. Nevertheless, every grunt always carried his loaded weapon with him and subconsciously located the closest foxhole or improvised fighting position...just in case.

A unit only stayed in a NDP for one or two nights. Then the men moved to a new location. The distance humped depended upon conditions—weather, terrain, and vegetation—and the mission. Movement generally was between one and ten kilometers, depending

---

103. Pronounced "oh five hundred."

104. Due to Vietnam's close proximity to the equator there was little variation in these times due to the season of the year.

upon terrain. In the steep mountains covered with thick tropical vegetation, even a single kilometer could be impossible. In the relatively flat region between the mountains and the South China Sea 10 to 20 kilometers (6.2 to 12.43 miles) was a walk in the park. A unit only had approximately 12 hours of daylight each day in which to move and prepare a new NDP. By 1800 hours (6:00 o'clock in the evening) night began to alter the landscape. The thick vegetation brought darkness to the jungle even earlier than elsewhere.

As soon as their unit stopped at a new site, the grunts dug foxholes. Then a sleeping position was set up. Finally, each man prepared an evening meal. He might have swallowed fruit, crackers, or candy during the hump, but his last cooked meal ordinarily had been breakfast. So now he was ready for a hot meal. His stove was an old C–Ration cracker can with triangular holes punched around the can's bottom and top rims using a "church key" bottle opener. Heat tablets were provided in the C–Rations to cook the meal. They were satisfactory for heating food. But C–4's bright blue flame was much hotter and better for boiling water for coffee or cocoa.

C-Ration packs contained matches for igniting the fuel. However, grunts needed a more reliable method in wet monsoon conditions. Therefore, in addition to matches every grunt carried a Zippo® cigarette lighter for starting all types of fire, from smoking cigarettes to incinerating enemy huts. My Zippo® had the 187th regimental crest attached to one side of the lighter. Above the crest on the cap was an engraved CIB. The other side had "PAM & LEBRON" engraved on the cap. The body was engraved "C-3/187" over "VIETNAM" over "70-71."

Everyone also carried a P38 can opener to open his meal. Some attached them to the chain of their dog tags. Others stuck them in the camouflage band on their helmet. The P38 had been invented during World War II and came in U.S. Army ration boxes until well past the Vietnam era. The P38 was about an inch and a half long. It had a small notch and a hinged blade on one side. The notch was placed over the rim of a can's end and the blade was flipped out perpendicular to the opener's body. This curved blade functioned like a claw. With a swift, back and forth motion of the wrist, grunts easily peeled away the lid of metal cans containing their favorite meal. Mealtime concluded with site sanitation. Empty cans and other garbage were buried in a shallow hole.

After eating, if a grunt was lucky, enough daylight remained for some leisure. Leisure activities included reading and writing letters, reading books, playing cards, and conversation. Paperback novels and Gideon Bibles were the most common reading materials. In Charlie Company, spades or hearts were the grunt's preferred card games. We kept score, but gambling was not customarily involved. Army gossip and postwar plans dominated talk. Discussions about the future ranged from how good the first Big Mac hamburger would taste to more serious topics, such as marriage and career. Some grunts pulled pictures of girlfriends and bragged about their beauty. One guy was a rodeo rider and he shared photos of bulls and horses he anticipated riding after the war.

Despite the diversity of backgrounds and difference of interests, the grunts in one's platoon were family, close family, close as brothers. They lived together and so shared their lives with each other. No matter how strange or boring a buddy's personal interests might seem to them, it became important to them because it was important to him. So a kid from New York City listened attentively to a Texan's raving about the rodeo. And a white kid from rural Alabama shared the cookies his mother baked with a black kid from New Orleans. Or a short-timer carried some of a stumbling cherry's burdensome gear. They did so because that's what brothers do.

Mostly grunts just tried to relax whenever given a chance to do so. The physical exertion, lack of sleep, harsh environment, and intense stress combined to intensify exhaustion, the grunt's constant companion. Early in my tour with the Rakkasans, I learned to live with this unwanted companion. I didn't like her, but once she caught you she never really let go. She consumed every fiber of your being. She constantly sought to introduce you to her colleagues, carelessness and feebleness. During every hump, you struggled against her in order to lift your foot and complete each step. She lacked the beauty of a Greek siren, but she was equally deadly, this companion of every grunt ... total exhaustion.

All light sources were extinguished after twilight. Even the faint red-orange tip of a lit cigarette could be seen from a distance. The only exception to this blackout was the olive green plastic flashlight that officers or their RTOs carried. This flashlight was eight inches long and had an angled head that projected the light beam at 90° from its body. A red plastic filter covered the lens. This permitted commanders to read maps and RTOs to operate their radios without destroying their night vision or casting a bright light in the darkness.

The army decided to extract Charlie Company on Thanksgiving Day for a special Thanksgiving meal. The menu included turkey, dressing, and other traditional holiday dishes. The extraction entailed the company moving several kilometers to a new LZ site. I still was relatively new to life in the bush and had developed some type of digestive ailment. That morning I was vomiting and suffered from a bad case of diarrhea. I still remember that day as my worst humping experience of the war. I was so weak from dehydration that I could barely stand up. It took every ounce of fortitude I possessed to complete the movement. I almost drowned crossing one swift mountain stream. Large stones created whitewater conditions and I slipped on the rock creek bed. The weight of my rucksack quickly pulled me under. I gasped for air but swallowed water. Fortunately, two grunts grabbed me and pulled up out of the water. Each one clutched a shoulder of my jacket and dragged me onto dry land. I was too sick to even know who my rescuers were. I recall my squad leader, Sergeant Jim Eggleston (known as Egg), checking on me after I got out of the water. So I'm fairly sure that he was one of my rescuers. Despite the magnificent feast that had been prepared, I ate only some dry toast and drank a glass of tea. Even so, the meal improved my condition, both mentally and physically.

After the meal, the company was supposed to return to the bush. As the crowd on the pad grew in number, a few grunts wandered around carrying on conversations with guys from other platoons. Others sought a short nap. Time passed without any birds showing up to transport us back into the field. Activity and conversation waned. Finally the battalion chaplain arrived and conducted a brief Thanksgiving service. It was one of the few opportunities I had for corporate worship while I was in Vietnam. Initially most of the guys were apathetic. Then about mid-way through the service, the NVA sent a few rockets into Evans, striking the main runway. The runway was far enough away that the blasts posed no danger to us, but they were close enough that the gray puffs of smoke were clearly visible. Their explosions transformed the men's spiritual demeanor wholly. Everyone seemed engaged in the rest of the service.

However, the hour was late, too late to make a company size combat assault. Charlie Company would stay at Evans that night. Therefore, the grunts drifted down to the company hooches. I went straight to bed. My body needed the sleep in order to recover fully from the stomach illness. Despite knowing we would be going back to the bush the next morning, many grunts consumed copious quantities of alcohol that night. Uncle Al recalled seeing grunts passed out on the parade ground and between buildings the next morning.[105]

---

105. Alan "Uncle Al" Davies, in conversations at the 101st Airborne Division

THE FLAG WAS STILL THERE

NCOs scurried about waking their charges at daylight. After a hot breakfast, the company assembled on the battalion helipad for our return to the bush. Across the pad, an oversized wood *torii* stood silhouetted against the blue sky, a silent sentinel over our gathering. Content from a full stomach, most grunts reclined against the back of their rucksack for a brief rest. Eyes inevitably were drawn to the giant red symbol of the Rakkasans. It reminded me that we were in Vietnam and so elicited contradictory subconscious emotions. Anger and disgust for being here rather than at home on Thanksgiving certainly dominated. Yet pride in belonging to this unit was also there. This was the battalion that took Hamburger Hill in May 1969. But the unit was not just some illustrious army organization. It was the men sitting around me, the men of Charlie Company, Third of the One-eight-seven Infantry, 101st Airborne Division. They considered themselves to be the best in the world. None of us knew what we would encounter back in the bush, but we knew we could depend on each other.

Rear echelon morale in Vietnam had deteriorated significantly by late 1970. Consequently, the quality of clerical work sometimes suffered. Paperwork sometimes became sloppy. I was relieved of duty with "Co A 1st Bn 12th Inf" [my company in the 4th Infantry Division] on 23 October 1970 and ordered to report to "101 AG Admin Co 101 ABN Div" [101st Airborne Division] on 5 November.[106] Orders reassigning me from "Det 1, 101 Admin Co (Airmobile)" [replacement depot] to "HHC, 3d Bn (Ambl) 187th Inf" [battalion headquarters] were dated 17 November 1970. Furthermore my last name was misspelled "MATTEWS."[107] Orders reassigning me to

---

Snowbird Reunion in Tampa, Florida, February 8-10, 2018.

106. Headquarters 4th Infantry Division, Special Orders Number 296, 23 October 1970.

107. Headquarters 101st Airborne Division (Airmobile), Special Orders Number

"Co C 3d Bn (Ambl) 187th Inf" [my combat company in the 101st Airborne Division] were not issued until 27 December 1970. By then I already had been in the field with the company at least six weeks. The misspelling of my last name was retained on the reassignment orders. In other words, I had experienced combat as an infantryman with Charlie Company long before I was assigned to it on paper.

The administrative laxity produced distressing consequences for Pamela, my parents, and the rest of my family. In the transfer from the 4th Infantry to the 101st Airborne certain records of my existence mysteriously disappeared. So I too officially ceased to exist. Pamela's account of the episode follows:

> *I wrote letters daily to LeBron and sometimes baked him cookies and mailed to him. His mail back to me came in bunches. I would miss several days and then get all the mail on one day. It had been a long time since I had received anything but I knew that when he was out in the thick of things the mail sometimes was not picked up. Time kept moving along and no letters until one day my postman, Mike Pass, who was an older brother to one of my friends at school, knocked on the door. Usually he just put the mail in the slot in the door and moved on but not on this day. I opened the door to him standing there with a month's worth of letters and a box that held peanut butter cookies. They were not letters for me but from me. Mike looked at me with a broken heart and said, "Pam, I am really sorry." He handed me the box and letters and walked away without another word. Stamped on each article was "Addressee Unknown." My first thought was MIA (Missing in Action) or even worse, dead. I burst into tears and ran to the phone to call LeBron's parents. They too had received letters back.*

321, 17 November 1970.

Although I am sure not as many as I had received because I wrote all the time. Ironically, I had just received an assignment in my [high school] Government class. It was to write my Senator about a topic that was important to me. I now had an important topic, LeBron. I wrote Senator Herman Talmadge asking him to find LeBron. Simultaneously LeBron's father took to the same mission of finding his son by contacting our Senator.[108]

Herman Talmadge (1913-2002) had been in the United States Senate since 1956. Prior to that, he twice had served as Georgia's governor (1947, 1948-1954).[109] In 1970 he was a very powerful politician in Washington. His office quickly started an investigation.

Meanwhile I trudged wearily through the monsoon, oblivious to the mail situation. In the bush we were totally dependent upon the UH-1D/H Huey helicopter for resupply. Once or twice a week a Huey was supposed to bring us C–Rations and other supplies. Most of the time, they also delivered a red or orange nylon mailbag. Mail was essential to morale. In those days it usually took five to seven days for a letter to go from the States to the field. All mail was addressed to soldiers via one's unit through APO San Francisco, California.[110] One of the few benefits for soldiers in Vietnam was not paying postage for letters we mailed. Instead of affixing a ten-cent air mail stamp, we simply wrote "free" in the upper right corner of the envelope.

Grunts in the bush relished the sight of a red mailbag when the resupply helicopter touched down. Virtually the entire unit anxiously awaited the distribution of its contents. I had been painfully

---

108. Pamela Coffman Matthews (June 11, 2013).

109. The New Georgia Encyclopedia (http://www.georgiaencyclopedia.org/nge/Article.jsp?id=h-590). Accessed August 4, 2017.

110. APO stands for Army Post Office.

disappointed when each bag was devoid of any mail for me. I assumed the problem was typical army red tape because of my transfer from the Fourth Division. Furthermore, the rainy weather interfered with helicopters flying and resupply often was sporadic. Consequently going without rations for a day or more was common. Hence, food often overshadowed other considerations.[111] But even food for an empty belly did not fully eliminate my anguish.

After weeks in the bush, Charlie Company returned to Camp Evans for stand-down. After eating a hot meal in the mess hall I picked up my rucksack and ambled down to the platoon hooch. A few minutes after stowing my gear away, the first sergeant summoned me to the company orderly room. The unexpected call made me dreadfully apprehensive; such calls rarely involved good news. *Was I in trouble?* I could not think of anything I had done wrong. So I began to speculate about tragedy at home. When I reported to the first sergeant, he handed me a small stack of letters and briefly interrogated me about when I last received mail. Next, he said the company commander wished to see me. I did not yet realize what was going on and so I remained very uneasy as I waited to go into the CO's office.[112] Following what seemed an eternity, the CO summoned me inside. I saluted nervously. He casually returned the salute. After putting me at ease, he looked me directly in the eye and explained the situation. Then he handed me a printed list of addresses in my chain of command, beginning with him and going up through the Army Chief of Staff and the Secretary of Defense. He then put me in a room by myself where I had to write a letter to every person on the list stating that I was alive and receiving my mail. I also had to write similar letters to Senator Talmadge, Pamela, and my parents. I wrote at least a dozen

---

111. Letter to my parents dated "22 Jan 71.
112. CO stands for Commanding Officer.

letters. Well essentially, except for Pamela and the family, I wrote one form letter. Then I copied it by hand, changing only the addressee. The ordeal was quite a hassle for me, but I never failed to get mail after that.

It was not until many years later that I began to understand the toll Vietnam exacted on family members of the men and women serving in that conflict. The sorrow Tommy's death caused Uncle Winfred and his family was obvious. Then again the families who did not experience causalities also suffered. My sister Lisa was nine years of age when I left for Vietnam. She described my deployment's impact on my parents:

> *I remember Mother and Daddy watching the news about what was going on over there, I remember they [the television reporters] reported on how many people were killed, and I remember sometimes they [Mother and Daddy] would get your letters and talk about how you were in some place that had been in the news weeks before but it took the letters so long to get here that the bad fighting was over by the time they knew you had been there.*[113]

Separation, anxiety about receiving that dreaded telegraph, and a host of other consequences placed an underappreciated strain on the grunt's family.

Race relations reached a low point among the Rakkasans during November and December 1970. Many African-Americans in the battalion organized themselves into a paramilitary fraternity under radical leadership. They occupied one hooch in the battalion area, which white soldiers derisively designated as "the Ghetto of the

---

113. Lisa M. Lewis, e-mail message to the author, October 8, 2013. Bracketed words inserted by the author for clarification.

Harlem Hooch." The strong odor of pot oozed out of the hooch and the loud notes of soul music blasted the space outside the hooch. Its occupants wore distinctive armbands and for the most part stayed in the rear. Many blatantly refused to go fight "the white man's war." White grunts resented their failure to join the company in the field. As Charlie Company lingered on the helipad to return to the bush after the December stand-down, the grunts on the helipad released their resentment in verbal accusations and derogatory invectives. Skin color meant nothing in the bush. However, the absence of the black malcontents threatened to destroy the credence of all black soldiers by their white comrades. Howard Dove was one of two or three black soldiers waiting to return to the bush. Dove was large, with an athletic build. Gossip claimed that he had played football at the University of Oklahoma. As the vindictive rhetoric increased, he quietly walked away carrying only his M-16 rifle. A short time later a half dozen more black soldiers reluctantly sauntered up in field gear. Dove walked behind them, his rifle leveled at their backs. It was locked and loaded. I believe he actually would have shot them if they had balked. Dove already was respected by all the grunts for his superb performance in the bush. That afternoon they gained an even greater appreciation for the black kid who quietly did his job. They knew if the company got into a tight spot, they could depend on Dove… and Dove could count on them.

The race problem came to a head while Charlie Company was in the field. During the Christmas holidays, the battalion executive officer entered the hooch drunk and verbally assaulted the black soldiers inside. The following night someone fragged an unoccupied officer's quarters. Colonel Sutton then acted to break up the Harlem Hooch. Rumor spread that he entered the quarters and someone pulled back the charging handle on a M-16. Whatever occurred, Sutton summoned the Military Police. They surrounded the hooch with armored security vehicles. The MPs then arrested those soldiers inside the hooch. Most of those not charged with major crimes were transferred out of the battalion. A few radical ringleaders were court-martialed and sent to the notorious LBJ (Long Binh Jail). Sutton also transferred the offensive executive officer out of the battalion. Race relations became less overtly antagonistic thereafter.

On December 22, 1970, Bob Hope arrived at Camp Eagle to perform his 20th Annual Christmas show for the 101st Airborne Division. Hope treated more than 19,000 troops to two and a half hours of singing and comedy. Among the featured performers were singer-dancer Lola Falana, country singer Bobbi Martin, singer Gloria Loring, the Golddiggers from the Dean Martin television show, the Ding-A-Lings, Les Brown and his "Band of Renown," Cincinnati Reds catcher Johnny Bench, and Miss World Jennifer Josephine Hosten.[114] Everyone who was privileged to attend professed it to be a fantastic show.

I was among those selected from Charlie Company to attend the show. Unfortunately Charlie Company was among the last units scheduled for pick up. For some reason, helicopters were not able to pick us up at the appointed time. Sometime later we received a report that officially we were "socked in," meaning weather conditions would not permit helicopters to fly where we were located. To this day, I believe the pilots did not want to be late to the show, and that was the real reason they did not pick us up. Visibility on the ridgeline where we were located was excellent. One could see for miles.

I cannot confirm that it was legitimate, but the grunts in Charlie Company understood a Christmas cease-fire was called for December 25. All units were scheduled to remain in place on Christmas.

---

114. Thomas Rutledge, "Screaming Eagle Vietnam Diary," *Rendezvous with Destiny* (Winter/Spring 1971), p. 35.

However, on Christmas Day Charlie Company ended up moving off a mountain near the location of FSB Ripcord. Russ Kagy slipped while climbing over a log. He grabbed a bamboo stalk to keep from sliding down the mountainside. The bamboo sliced open his hand between his thumb and forefinger. Doc was bandaging the wound when the point team set off a booby trap. Three guys—including Howard Dove—were seriously wounded and required a Dust-Off medevac.

Medevac is an abbreviated term for medical evacuation. Medevacs in Vietnam employed Huey helicopters fitted as air ambulances to evacuate wounded soldiers off the battlefield. In World War II, on average it took 10 hours and 30 minutes to move a wounded G.I. from the battlefield to hospital treatment. In Korea, the time was cut to 6 hours and 20 minutes. In Vietnam, it had been reduced to 2 hours and 48 minutes. Many soldiers reached the hospital within 20 minutes of being injured.[115]

Booby traps were a constant danger in the bush. In 1970, 80% of all American wounds were caused by mines or booby traps.[116] The most common device encountered by Charlie Company in late 1970 and early 1971 was a quarter pound of C-4 plastic explosive with a pressure detonating blasting cap. The resultant explosion would shatter the foot and lower leg, necessitating traumatic amputation just below the knee. The Christmas Day device was a mortar or artillery round detonated with a trip wire. It was a cruel reminder that this Christmas there was no peace on earth, goodwill to men.

---

115. *The Encyclopedia of the Vietnam War*, 2000 ed., s.v. "Medicine, Military" by Jack McCallum, p. 261.
116. *Ibid.*, p. 261.

JANUARY 1971

1971 Brings Heavier Combat[117]

A NEW YEAR

Right after Christmas, Charlie Company returned to Camp Evans for stand-down. I had been in Vietnam for over three months and seen little actual combat. I had been in a few fleeting firefights with undersized NVA squads, but nothing significant. The monsoon weather had been my worst enemy. Bad weather again set in and the standdown was extended a couple of extra days. Initially everyone was delighted. Then boredom bred trouble. Young men trained in a violent occupation plus the easy availability of alcohol and other drugs was the formula for conflict. Several fist fights occurred. The brotherhood forged in the bush was breaking down. Finally late in the afternoon of the ninth day, the company returned to the bush. Due to a shortage of available helicopters, the insertion required several trips.

---

117. Letter to my parents dated "20 Jan 71"—The 12 page letter documented in detail my role in the Falls operation. I debated sending it for two days before writing a cover letter and mailing both together [letter to my parents dated "22 Jan 71"].

***

The Assault Helicopter companies of the 101st Airborne Division operated the Bell "Huey" UH-1H model at this time.[118] Its official name was Iroquois but the nickname Huey became its iconic identity. "Huey" was derived from the helicopter's original designation, HU-1 (Helicopter, Utility Model 1).[119] Grunts also referred to them as "birds," "choppers," and "slicks." Birds obviously pointed to the vehicles flying function. Chopper seems to be due to the sound of the rotary blade's action. They were known as slicks because of the uncluttered external appearance when compared to earlier helicopter designs. The first Huey entered service in 1959. The improved UH-1D was delivered to the army in 1963. It was 57 feet, 1 inch long. The UH-1H was introduced in 1967. Externally it was virtually identical to the UH-1D, but was fitted with a more powerful engine. By 1971 all UH-1D models had been modified into UH-1H models. The Huey revolutionized warfare. The parachute had introduced a third dimension to the battlefield, airborne assault. The helicopter transformed airborne assault from a preparatory raid preceding a more conventional strike by land or sea into a potential principal attack route. In Vietnam, the helicopter CA (Combat Assault) made this third dimension the primary delivery means.

The helicopter CA was integral to American ground tactics and strategy in Vietnam. The acronym CA applied to any helicopter movement of ground forces into hostile territory. The helicopter could rapidly carry infantry over long distances and deliver them with all their equipment to any point on the battlefield. Depending upon

---

118. Vietnam Ghost Riders (Co. A, 158th Aviation Bn.) website (http://ghostriders-online.org/huey_specs.html). Accessed July 3, 2017.

119. Dunstan, *Vietnam Choppers*, p. 20, n. 1.

the size of the assault force and the terrain, helicopters conducting a CA employed a variety of formations. The most common was the "V" formation in which Huey HU-1H helicopters flew in echelon. The four-bird diamond formation also was employed regularly. The formation flew at a high altitude during the trip to the LZ in order to avoid ground fire. Terrain and weather factored into determining exact altitude. Just prior to reaching the LZ, the helicopters descended swiftly to a low altitude. They then approached the LZ at high speed, just above the treetops.

A preparatory barrage struck the LZ for five to ten minutes before the first helicopter touched down. The barrages employed air strikes and/or heavy artillery. Firing stopped one to two minutes before the first troop transport landed. The last round in the barrage always was white phosphorous. White phosphorous, Willie Pete as it was known in army radio jargon, exploded with a distinctive white smoke outburst. This signature burst notified the first slick that from the perspective of friendly fire it was safe to enter the LZ.[120] About 30 seconds before the transport helicopters touched down, AH-1G Cobra helicopter gunships opened fire on the LZ.[121] They ceased fire and pulled away as the Hueys made their final approach. During the CA, the Cobras circled the LZ ready to provide suppressive fire if the troops on the ground encountered enemy resistance. Door gunners on the slicks also sprayed the jungle around the LZ as the grunts jumped from the Hueys and raced to their preassigned position on the perimeter of the LZ. If there was no enemy present, the LZ was labeled "cold." If the CA did encounter enemy opposition, the LZ was said to be "hot." The majority of CAs were cold, but the status usually remained unknown until seconds before the first boots touched the ground.

---

120. *Ibid.*, 65; for an aviation unit's perspective, see pp. 60-73.
121. *Ibid.*, p. 115.

One UH-1H normally transported five or six fully equipped American grunts when making a CA. Only the crew rode in seats. The grunts sat on the floor of the Huey, two on each side with their feet dangling free outside the cabin. The fifth and sixth grunts sat in the center of the cargo space floor. Before the helicopter touched down, the four men in the doorways stood up on the skids. The fifth and sixth sat up and crouched facing opposite doors, each ready to spring out as soon as his two comrades on the floor were clear of the opening. The helicopters rarely landed. Normally the chopper hovered about two feet above the ground. It stayed there only briefly and then quickly hurried away. Staying too long drew enemy fire, risking both grunts and aircraft crew. As the first wave set down, the grunts jumped out of the helicopters onto the LZ. Hunching over to avoid being hit by the main rotary blade, they sprinted out to the edge of the LZ amidst the dust and debris stirred up by the prop wash. Taking advantage of whatever cover was available, the grunts quickly began to establish a defensive perimeter to secure the LZ. Speed was a critical factor in these operations. Once the first Huey discharged its troops, masking the LZ's location from the enemy became impossible. Therefore, operation planners and grunts on the ground presumed the enemy would move to disrupt the CA and destroy the landing force before it could organize a solid defense.

Ideally all the helicopters in a CA would set down on the LZ simultaneously. Unfortunately, in most cases, particularly in the mountainous terrain in Charlie Company's AO, this was not possible. Therefore, the first grunts on the ground were alone until a second wave of helicopters set down. The fewer choppers in each wave, the longer the landing force's vulnerability lasted. Frequently only one or two helicopters could land at a time because of thick jungle or steep slopes. That meant for a two-minute interval five to ten men were isolated on the ground and threatened by an enemy force of unknown size and proximity. The capabilities of the Huey, its crew, and the grunts it carried automatically determined the length of this timespan. It typically took a helicopter two minutes to land troops: thirty seconds on the approach, one minute to unload the grunts, and thirty seconds to exit the LZ.[122] Two minutes may not be long, but in combat, two minutes can be the difference in life or death. Hence, for the initial wave of grunts, the few minutes between landings seemed like an eternity. Establishing a complete perimeter around the LZ with only ten men was impossible. Instead, the limited number of grunts on the ground set up a small defensive position to secure their personal safety. As additional troops landed, the defensive position expanded along the periphery of the LZ. In due course, a circle of combat infantrymen surrounded the LZ. Offensive operations only commenced after the unit's last man had been delivered. Until then the mission remained defensive, to secure the LZ for the remaining troops to land.

* * *

An early-January break in the weather enabled a significant increase in air support capabilities. Consequently, both Bravo and Charlie Companies from the 3rd of 187th Infantry (Airmobile)[123] were inserted along the enemy supply route through the Falls AO. The Falls AO was somewhere in the high mountains east of and adjacent to the A Shau Valley. The terrain was extremely steep and covered predominantly with double and triple canopy jungle. The Song Bo and its tributaries offered the NVA a convenient infiltration means from the A Shau into the coastal plains north of Hue. The mission called for

---

122. *Ibid.*, p. 65.
123. Hereinafter cited as 3/187 Infantry.

the two Rakkasan companies to interdict and terminate this enemy access route to the civilian population living along the coastal plain.

The January CA into the Falls AO occurred in disjointed waves. Battalion scheduled Bravo Company to go in first. Charlie Company was still on their extended stand down at Camp Evans. Bravo Company was still in the bush and theoretically its PZ was closer to the chosen LZ. However, for some reason only two choppers were originally allocated to the task. That meant the initial 10 or 12 men would be alone on the LZ much longer than two minutes. These grunts would have to wait alone while the two choppers flew back to Bravo Company's PZ, loaded the next 10 or 12 men, and then returned to the LZ. Naturally, implementation of the plan immediately went awry. After landing only ten men at the designated LZ, the meteorological conditions changed rapidly at Bravo Company's PZ. Incoming choppers were unable to land due to the inclement weather. Consequently, completion of the initial CA shifted to Charlie Company. The logistics of this change took even more time. The longer the transfer took, the greater the danger escalated for the ten isolated grunts on the LZ. After landing Charlie Company's first wave on the LZ, the helicopters returned to Camp Evans and picked up the second wave. Those grunts around the LZ tensely waited for the choppers to return. Their ability to defend the foothold was tentative without reinforcements. Each subsequent relay added more firepower to their military capability.

As soon as Charlie Company established a secure perimeter around the LZ, Captain Turk, the company's CO, sent first squad from first platoon out on a RIF. RIF was the Rakkasans' customary term for a combat patrol. It was an acronym for Reconnaissance in Force. Reconnaissance in force? *In force* is relative. A single squad hardly qualified as "in force" in this new AO. The mountains on the far side of the valley into which we were descending formed the eastern border of the infamous A Shau Valley, a fact that did not register on me fully at the time. After being in the company over two months, I had become accustomed to life in the bush. It was now my normal. Accordingly I did not question sending ten men out on a patrol. In hindsight, I recognize now that we were in an area where the NVA operated in larger formations than they did in the places where I had been to date. At the time, a certain amount of naïve enthusiasm gripped me. I dropped my rucksack and helmet with those of the other members of the RIF. The patrol replaced their discarded helmets with boonie caps. I pulled mine out from my rucksack and did likewise. A bandoleer of M-60 ammo was slung across my chest. I checked the canteen on my belt to insure it was full of water and formed up behind Egg's RTO (radioman[124]). Kevin Owen took point, his trademark boonie cap perched perfectly on his head. The crown was flat but the brim was curved down over each ear like a Spanish conquistador's helmet. Cowboy walked slack. Egg and his RTO were third in line. The rest of squad fell in behind me as we quietly slipped out of the perimeter and moved cautiously down into the valley.

The RIF went out about 200 meters from the LZ and then began to circle back by a different route. Suddenly the RIF encountered a "*highspeed* [sic.] *NVA trail.*"[125] We cautiously scouted down the trail and soon discovered footprints in the mud. The condition of the tracks after the recent rain suggested they were made sometime that same day. Not enough footprints were legible to determine the number of enemy soldiers involved. Egg duly reported the information to the CO by radio. Thereafter the squad made its way back to the LZ. We were fortunate our RIF that morning did not encounter a sizable

---

124. RTO was an acronym for Radio Telephone Operator. In Vietnam an infantry RTO only carried a radio. The field telephone had become obsolete.
125. Letter to my parents dated "20 Jan 71."

hostile force. Before the month was out, however, we would gain an opportunity to measure ourselves against these larger NVA units.

Later that afternoon, the entire first platoon moved out to probe toward the river. We started down the same route as the morning RIF. The footprints found on the earlier patrol made everyone keenly aware that the enemy was nearby. No one knew the strength or type force. Was it NVA or VC? Was it only a small squad or a full division? We moved cautiously down the steep hillside, moving closer to the A Shau, watching for any sign of the foe. By the time that first platoon moved about 300 meters, dark shadows started to obscure the landscape. The distance back to the LZ was short but conditions precluded night travel. The topography was near vertical slopes covered in dense jungle vegetation. Visibility was zero in the dark. We would wait for daylight before rejoining the company.

Our solitary platoon set up a quick NDP. Digging foxholes was not feasible. The ground was too hard and the sound of digging would disclose our presence. So each man took cover behind trees or other types of natural concealment and waited nervously for the first glimpse of the sun. The evening meal consisted of whatever cold can of C-Rations we could dig out of our rucksacks. Air mattresses and poncho liners remained packed. That night no one would sleep much. Officially, one in two men was on guard duty at any given time. The other man tried to snooze, but the platoon's isolation made slumber difficult.

The entire platoon was awake before dawn. Everyone peered intently into the darkness. The minutes just before sunrise always seemed the most dangerous. The dim light shrouded reality in the tropical rain forest. The dark grey shadows masked the presence of an adversary. Conversely, the limited light altered plants into hordes of Communist soldiers. Tense hands clutched M-16 and M-60 pistol grips. Fingers rested snugly against the trigger, ready to squeeze at the slightest provocation. The sun rose that morning without incident. First platoon packed up and climbed the steep hillside back toward the LZ.

The rest of Bravo Company managed to make it into the LZ while first platoon was away on its solitary RIF. The last sorties of Bravo did not touch down until after nightfall. All the grunts of both companies finally were together on the ground. However, the hour was too late to move to a stronger defensive position. As a result, the two companies (minus the first platoon of Charlie Company) remained on the LZ together that first night and a hasty NDP was established in the darkness. The late hour hampered digging proper foxholes for either company. Nonetheless, a few grunts scraped shallow fighting positions into the dirt in some places. Other grunts used fallen trees or rocks for cover and concealment.

It was still early morning as first platoon approached the crest of the ridgeline. The grunts of Bravo and Charlie Companies on the LZ ate their C-Rations and packed their gear. Around their perimeter, sentries—the two companies already had lowered their alert from 100% to 50%—gazed out into the shadowy jungle in a hopeless attempt to penetrate its curtain and see the foe they imagined was there. Dark shadows still bamboozled these sentinels into imagining an enemy force approached undetected. They also were still confused and jittery after the previous day's mishaps. Consequently, they likely would shoot at the first sound from outside their NDP. Friendly fire can kill you as quickly as hostile fire.

And so a short distance from the company perimeter first platoon paused to wait while Lieutenant Smith, first platoon's platoon leader, contacted the company CP over the radio. Then the CP passed the word to the grunts on top of the hill of our approach. Thereafter we moved up the hill and entered the safety of their NDP. They looked at us with envy and relief... envy because we were returning safely from danger they knew they soon must face. Next RIF into the enemy's lair might be theirs and it might not end as placid. Yet there was relief because our return increased their firepower and thereby improved the company's chances of survival when the inevitable encounter with the NVA occurred. More notably, we were brothers and we were back with the extended family safely.

When we arrived, Captain Turk sent out second and third platoons in search of the elusive enemy. Turk and his RTO remained on the LZ. Lieutenant Smith put me in command of three other grunts and assigned us the task of securing the LZ. Four men hardly could defend the assigned position against any serious NVA attack. At the time I was too proud of being assigned the responsibility of command to realize the lieutenant likely considered us the four newest and most expendable men in the platoon at the time. Because I was an in-country transfer, I had seniority over the other three and therefore was put in charge. Our mission simply was to prevent another incident like occurred the previous year when the three VC rifled our abandoned NDP. The rest of first platoon set out to locate a better NDP. It would serve as a temporary tactical defensive position from which Bravo and Charlie Companies could launch combined offensive operations.

While second platoon was on their RIF, it spotted at least five NVA soldiers — probably trail watchers — and gave chase. The five dinks[126] attempted to improvise an ambush of their pursuers, but their strategy miscarried. Second platoon killed one of them in a brief firefight. The other four escaped. Late that afternoon, second and third platoons returned to the LZ from their RIFs. First platoon had established a new NDP nearby. The entire company assembled

---

126. "Dinks" was Charlie Company's favored derogatory slang designation for the NVA and VC. Elsewhere in Vietnam, "gooks" was the more common term for anyone of Asian origin.

there and set up for the night. Bravo Company established a separate NDP nearby.

The next day's instructions called for first platoon to move another 400 meters, set up a perimeter, and then send out a couple of RIFs. Just short of reaching our objective, first platoon received new orders. We were to link back up with the rest of our company and move jointly with Bravo Company against a suspected NVA village.

A Recon team had travelled through this area a few weeks earlier. The team made contact with the NVA three times before they could pull out of the area. Recon teams were small, specialized units assigned to a reconnaissance role. They commonly operated in enemy territories, gathering intelligence or conducting small-scale raids. They lacked sufficient personnel and firepower for direct engagement with large enemy units. This team's contacts and its subsequent withdrawal signaled sizable NVA units operated freely in the area. These enemy forces deemed the Falls AO their territory. They were unlikely to allow us unimpeded access or to cede control of it to us without a fight. My tour of duty in Vietnam was turning into a real war.

The mountain range was rugged, steep, and covered primarily in triple canopy jungle. After moving 500 meters, we received a report that Bravo Company had hit some booby-traps. We then humped another 900 meters before linking up with the rest of Charlie Company. According to the map, the distance traveled that day was extremely short, only slightly over a mile.[127] However, the terrain was exceptionally rough. The grade was particularly steep, often the slope being 60 degrees or better. Heavy underbrush covered many of these hillsides. So trails had to be cut with machetes. Moreover, each man carried on average between 70 to 90 pounds of equipment. Consequently, it

---

127. One meter equals 1.094 yards. There are 1,760 yards in a mile. Therefore 1800 meters equal 1.118 miles.

was late when first platoon arrived at the company NDP. Despite the short distance covered, all of us were exhausted.

Bravo Company entered the NDP at the same time as first platoon. Originally, they intended to establish an independent NDP on a hill opposite the suspected village from our company's NDP. The proposed site would have situated American troops on two sides of the alleged enemy location. After taking casualties from two booby traps, Bravo Company's grunts were somewhat dazed. So they fell back on Charlie Company instead of establishing a separate NDP.

The booby traps they struck were similar to those Charlie Company encountered on Christmas Day, artillery shells or mortar rounds detonated with a trip line. The trip line was fabricated from adjacent vines, making it virtually impossible to distinguish.

Bravo Company lost one man killed, a medic named Stephen Paul Krug from San Mateo, California. Krug had been in country six months and was three days short of his twenty-second birthday.[128] Five other men were wounded (one seriously).

Our arrival, along with that of Bravo Company, required reconfiguring the existing NDP. More than doubling the number of troops formerly camped in the location expanded the NDP's perimeter significantly. The presence of two full infantry companies enabled the commanders to solidify the defensive positions on the perimeter and to extend defensive measures beyond the NDP perimeter. Because Bravo had suffered causalities that day, Charlie Company carried out most of these external measures.

Second platoon from Charlie Company set up an ambush as a security measure. Any NVA in the village certainly must be aware of the American presence in the area. The ambush was placed 250

---

128. The Virtual Wall ® Vietnam Veterans Memorial (http://www.virtualwall.org/dk/KrugSP01a.htm). Accessed July 4, 2017.

meters from our NDP in the direction of the suspected village. Hopefully, the ambush would thwart any NVA attacks from the village. My squad set up an OP (Observation Post) about 50 meters on the western slope below the NDP. An OP was a small position set up as an early warning for its parent unit's defense. Other platoons also set up additional OPs to cover enemy advances from other directions.

The jungle at night was a terrifying environment in its own right. It impacts all of one's human senses. The humid air is scented with rotting vegetation and mildew. Overhead foliage blocks all light from the moon and stars. Vision in the pitch-blackness often is restricted to a few inches at best. Fungi on decaying plants gave off an eerie glow in scattered locations. This glow's limited radiance failed to illuminate the dark. It only added to the unnerving atmosphere. The jungle's indigenous nocturnal animals played a symphony of unsettling sounds. Their performance took no intermission.

Add to all this ecological intimidation the knowledge that in close proximity were an unknown number of enemy soldiers who wanted to kill you. In this exotic jungle you were in their backyard. It was unfamiliar and mysterious to you, but they had lived here for years. Located outside the NDP perimeter, grunts in an OP often felt virtually isolated and alone. Quickly, duty in an OP percolated into a most intimidating experience. The slightest modification of sound or wisp of wind conjured up hosts of NVA thugs slipping up to slit one's throat. Hypothetically, half the men in an OP were supposed to be alert at all times. Even this was problematic. In reality, only sheer exhaustion made any sleep possible. We could not talk with each other to ease the tension. The enemy might hear us and pinpoint our location. As a result, there was little slumber for anyone that night… just endless hours of anxiety. I was greatly relieved when the OPs were pulled in the next morning.

Shortly after dawn, Bravo Company moved out to attack the

suspected NVA village. All civilians were cleared out of the area several years earlier. Therefore, any village found in this area unquestionably was an NVA depot of some type. Charlie Company's primary role was to act as a blocking force to prevent any NVA from escaping during the attack. Four ARVN companies also moved into positions to block other NVA avenues of retreat. As soon as Bravo Company passed out of the NDP perimeter, Charlie Company dispersed its three platoons to fulfill its mission in the day's operation. Second platoon moved out parallel with Bravo Company, securing its flank. Third platoon moved down to the river in order to prevent the NVA from using it as a means of fleeing Bravo Company's approach.

Both Bravo and Charlie Companies left their rucksacks stacked inside the NDP when they left on their assigned missions. Captain Turk assigned the task of securing both company's equipment to first platoon. Since a single platoon could not occupy the extensive perimeter of two full companies, Lieutenant Smith consolidated the platoon into a much more compact perimeter. He also sent out Larry Klag, Joe Sej, and me as an OP. We set up in the same spot that had been second platoon's ambush site the previous night. The circumstances precluded much conversation, but occasions such as this intensified the bond that exists between members of an infantry platoon.

Bravo Company found numerous bunkers and a few thatched hooches at the site of the suspected village. The bunkers and hooches contained significant quantities of food and military supplies. Bravo Company blew up the bunkers with C-4 and burned the hooches, destroying their contents. Conversely, they only encountered two enemy soldiers in the village. They shot and killed one. The other one escaped. Second platoon also discovered several booby-traps and a number of bunkers nearby. They detonated the booby-traps and destroyed the bunkers, but encountered no enemy soldiers.

After blowing up the booby-traps and bunkers, second platoon pulled back to the NDP and rejoined first platoon. Third platoon also returned from the river, reuniting the company. The company spent the next couple of hours clearing a small LZ. A resupply bird landed late that afternoon. It brought ammunition to replace what had been expended, a week's allocation of C–Rations, dry socks, and mail. A detail of four grunts unloaded the bird quickly. The bird lifted up and returned to the relative safety of Camp Evans. The supplies were divided into two stacks, one for Bravo Company and one for Charlie Company.

Once we distributed our portion of these provisions, Charlie Company and Bravo Company swapped positions. The next day Charlie anticipated assuming the assault role. First platoon was the initial platoon to move out and exchange positions with Bravo. Bravo only had advanced about 600 meters during the day's operation. Consequently, we promptly located a platoon from Bravo Company and moved into its position. The relieved platoon left and soon reoccupied our old position. By the time the whole exchange movement was completed, it was almost dark. Charlie Company set up a hasty NDP and waited for daylight in its advanced location. We did not know it at the time, but for the rest of the operation Charlie Company would function independently of Bravo Company.

The next day, first platoon again secured the company perimeter as second and third platoons continued to sweep the area. Third platoon located a number of hooches with bunkers hidden beneath the huts. One contained an AK-47 and some explosives. Second platoon discovered another trail with tracks less than a few hours old. The enemy in the area remained elusive. To date, Bravo and Charlie Companies had killed only two NVA soldiers in the operation. These were hardly the large numbers of enemy casualties expected from the operation. On the other hand, one American had been killed and five had been wounded. The number of NVA wounded was unknown.

From a grunt's perspective, it seemed as if we did a lot of the proverbial walking in circles and accomplished nothing but exhaustion.

The weather during the operation generally had been beautiful. Only on the day that we made the longest hump had it been foul. The magnificent weather persisted as second and third platoons returned from their sweep. The sky was clear and the temperature was mild. That afternoon Charlie Company trekked down to a nearby stream. A secure perimeter was set up and each platoon took a turn in the cool mountain water. We washed, shaved, and filled our canteens. Afterwards everyone felt better. Moving back up the hillside, the company set up a strong NDP.

The next morning, first platoon moved out on a RIF. We came across more fresh tracks and bunkers. The platoon traveled about 800 meters until we reached the river. Most of the way we followed an NVA trail. Having reached our objective, the patrol returned to the company NDP. Details were assigned after chow to destroy the bunkers uncovered in two days of searching the area. I was part of the detail that destroyed the huts and bunkers third platoon discovered the previous day. The huts were well-built thatched structures. They were very large. A strong bunker was dug into the ground underneath each one. Around the huts were baskets of food, NVA fatigues, and other supplies. We also found a bag with five sticks of explosives used in making booby traps. This latter discovery was the most rewarding. I wrote my parents:

That is five guys who will go home with feet. I was kinda [sic.] *proud of myself.*[129]

On January 13, 1971, second platoon unearthed a mass grave containing the dead bodies of 14 enemy soldiers. They had been dead less than three weeks. Although the army tallied these corpses in our

---

129. Letter to my parents dated "20 Jan 71."

company's body count, we were not certain that our company had killed them. They more likely died during an air strike or from artillery fire.

The following morning the entire company packed up at first light. After eating a hasty breakfast from our dwindling supply of C-Rations, Charlie Company moved out for a new location. We humped all day, but only covered 600 meters due to the terrain. The mountain slope was extremely steep and covered in thick vegetation that required cutting with machetes. Movement was a slow, arduous process. Late in the afternoon, the company set up a NDP on a nearby hilltop. The hill was devoid of large trees. Thus it easily was cleared to serve as an LZ. Foxholes were dug around its perimeter. Each of the three platoons was responsible for a third of the perimeter. The three platoons rotated doing the tasks of cutting the LZ, digging their foxholes, and pulling security for the position. The work consumed the remainder of the daylight hours. As the last flicker of sunlight faded over the western horizon, grunts blew up their air mattresses, finalized guard duty assignments, and swallowed a last bite of cold C-Rations. It promised to be another long night in the bush.

Both second and third platoons sent out short RIFs early in the morning of January 15. One of the patrols was able to make a pathway down to the river from the hilltop. First platoon once more remained in the NDP to secure the LZ for resupply. The familiar whoop whoop sound of a Huey reverberated over the mountaintop before noon. A colored smoke grenade was popped. Pastel yellow smoke rose over the LZ only to be blown here and yon by the prop wash of the Huey. The wash also blew dirt and dust into the faces of those close to its descent. The Huey hovered a few inches off the ground. Grunts bent over and rushed to each side of the chopper. They reached into the bird's belly, grabbed its contents, and slung them to the ground. The bird quickly lifted up and flew away. Boxes of C-Rations lay scattered on the ground. In addition to a fresh supply of C–Rations, on this occasion we received hot chow. The meal was delivered in insulated olive green metal containers. And, of course, the resupply bird also brought a precious red bag with mail. In the afternoon I went down to the river to bathe and swim. The relaxed day definitely was a boost to our morale. In writing about it five days later, I said, "*Boy did I feel good!*"[130]

I frequently walked point during this timeframe. Therefore, I acquired a 12-guage pump shotgun during this resupply. Another member of the platoon carried it previously. His DEROS was only ten days away. So we swapped weapons. He grabbed my M-16 and hopped on board the chopper to go home. I took the shotgun and inspected it carefully. It was a commercial Remington 12-guage pump loaded with buckshot. At that moment I was unduly excited about acquiring it. I would soon regret the exchange.

The next morning we again moved to a new location. This time we hiked approximately "*a click*" (one kilometer), moving half the way down the side of the mountain. It was late before we located a tolerable site for a NDP. "Tolerable" did not mean adequate. The spot was not level. The slope was so steep that everyone slept with his feet pointed downhill. We drove wooden stakes in the ground at the foot of our air mattresses to prevent our bedrolls from sliding down the mountain. Despite the long, arduous hump and the poor sleeping positions, everyone was in good spirits. The company hoped to be extracted and return to Camp Evans for a short stand-down the next day.

The Falls Operation had been abnormally brief. Charlie Company customarily stayed out in the bush six to eight weeks between stand-downs. This time we had been in the field about two weeks. We

---

130. Letter to my parents dated "20 Jan 71."

had killed a couple of Dinks and destroyed some caches. However, mostly we seemed to walk in circles in hellish mountains, dig lots of foxholes, and climb even steeper mountains. To the average grunt, all that we accomplished was checking time off one's tour in Nam. *"Tic toc! Tic toc!" ticked the clock.* Everyone was three weeks closer to his DEROS and finishing his tour in the Nam. Grunts considered this to be a positive achievement.

THE CLIMAX OF THE FALLS OPERATION

January 17 began as a normal day in the boonies. We grunts were looking forward to stand-down and hoping for seven days at Camp Evans. In preparation for the extraction, Charlie Company moved down into the narrow valley below our NDP. The mountains on both sides of the valley were exceptionally steep, near vertical at many places. Fortunately, the recent good weather had dried the ground and made footing less treacherous. A swift whitewater river cut through the valley floor. Stone boulders and ancient landslides formed some traversable ground at the base of the slopes. The rugged terrain made climbing from the previous night's NDP to a mountaintop impractical. The ascent to the mountain crest would require at least a full day of humping. Battalion wanted us back at Evans before sunset. A small island downstream in the river offered the quickest destination for cutting a suitable LZ. So the company trudged down to the river and followed it toward the island. The point team spotted a cave on the far side of the river a few meters before reaching the island. The tip of the island was clearly visible from both sides of the river in front of the cave. One of the RIFs sent out the day before had reported the cave was an NVA bunker that contained rice and other food supplies. They destroyed the contents, but lacked the resources to demolish the rock cavern. Captain Turk, Charlie Company CO, halted the company and sent a party across the river to investigate the cave again. They reported that the cave had not been touched since yesterday's RIF.

Charlie Company's grunts dropped their heavy rucksacks along the riverbank. Turk made dispositions for the company to be extracted. One squad from each of the company's three platoons was assigned to the work party for cutting an LZ on the island. Two squads from first and third platoon set up security for the other squads' rucksacks and the company CP. CP was the acronym for Command Post, a unit's command and communication center. At the company level, in the field it usually accommodated the commanding officer and two RTOs. Frequently it also contained an artillery FO (Forward Observer) and his RTO. The job of these two artillerymen was to direct artillery fire support for the company. Turk sent second platoon's remaining squads downstream on a RIF. The work detail and the patrol moved out quietly.

Lieutenant Smith assigned me, along with the other members of first platoon's first squad, to the detail cutting the LZ. Scattered small trees and sparse undergrowth covered the rocky island. Rapid swings with machetes easily cleared the underbrush. However, the bigger trees had to be felled with C-4 plastic explosive. Several sticks of the explosives were tied to the trunk and det cord attached. Then the cry, "Fire in the hole!" echoed across the island. A loud explosion followed. The blast sent a tree crashing to the ground. Slowly the LZ was taking shape. In a couple of hours it would be ready to receive birds. Unfortunately, this was not to be.

The company CP lost radio contact with second platoon's RIF during the time the detail was cutting the LZ. When communications were reestablished, the CO learned that the RIF had encountered NVA. The patrol reported one wounded. The man had bullet holes in his hand and leg. Because of his injuries, his squad rigged a stretcher using a poncho and two rods cut from small trees. They currently were almost two and half kilometers downstream from the company. They had broken off the engagement and were working their way back to the company. Enemy casualties were unknown. The patrol asked for a Dust-Off medevac to extract their wounded soldier.

Work on the LZ continued with a new sense of urgency. Around 1500 hours (3:00 o'clock in the afternoon), we heard firing from the company area upstream. The distinctive sounds of AK-47 and the familiar M-16 were clearly audible. The volume of firing grew in intensity. The M-60 machine gun added its deep voice to the reverberation. We formed a defensive line across the island under the careful leadership of the NCOs. Grunts dropped into the prone firing position. Each man used whatever cover he could find, but little was available on the flat island. Thumbs clicked the selector lever from "safe" to "semi." Nervous fingers pressed gingerly against triggers. Everyone listened carefully to the firefight upstream. The AK fire clearly was coming from the far side of the river and the M-16 sounded like it was coming from the first platoon's position along the riverbank. However, the noise of M-79 grenades was absent.

The M-79 was a 40mm grenade launcher, nicknamed "the Thumper." It was a breech-loading weapon with a range of approximately 400 yards. The M-79 was very accurate. Charlie Company's grenadiers primarily fired high explosive and canister rounds.[131] The Thumper provided an infantry squad with precise short-range projectile firepower. The M-203 had replaced the M-79 in the grunt's arsenal by 1971. It had a 40mm grenade launcher mounted underneath an M-16. It was awkward to fire and its performance did not match

---

131. Gas and smoke grenades also were available but rarely used. Most grenadiers carried a couple of red smoke grenades to mark enemy targets for close air support.

the M-79. Nevertheless, it was the weapon that was available to us on the island that afternoon.

First platoon's second and third squads remained with the company CP when the work detail departed. Russ Kagy set up his M-60 in the elephant grass.[132] He aimed the machine gun down the trail that we just had used coming to the site. Lieutenant Smith and Lieutenant Sweetland walked by, heading for the river. Just as they passed out of his sight, Kagy spotted a long line of pith helmets moving slowly in the grass down the slope on the far side of the river. Before he could alarm anyone, a firefight erupted. The guys back around the company CP [see Figure 1] were lounging around, taking it easy when Hector Torres (nicknamed Recon) spotted a single NVA soldier entering the cave [bunker] across the river. He shouted the alarm and the grunts nearby grabbed weapons with the intention of blowing the lone dink away. However, by this time 15 to 20 NVA soldiers already were visible around the cave. Both sides opened up. Due to the steep slope of the adjacent ridgelines, both sides had their opponents pinned down, unable to move. The NVA position initially was superior in that the cave provided better cover. Charlie Company was exposed on rocky, open ground. Firing therefore intensified as each side attempted to suppress the other side's firepower.

Kagy had swung his 60 around and was pouring rounds into the enemy. He glanced up to see First Sergeant Dayton Herrington calmly standing beside him.

"What can I do?" asked Herrington.

"Get me more ammo."

Herrington immediately headed for the rucksacks lying where

---

132. Russ Kagy related what he witnessed in a telephone conversation of Monday, July 11, 2016. He, Alan Davies, and I discussed in more detail during the 101st Airborne Division Snowbird Reunion in 2018.

members of the cutting party and RIF had left them. Most contained at least one belt of ammunition for the valuable platoon machine guns. Each metal belt held 100 rounds. Herrington grabbed an arm full and deposited them with Kagy. Russ's M-60 continued its sweet cadence uninterrupted.

We had two M-203s on the island. One of the M-203 gunners suggested, "Let's crawl up to the end of the island and lob some M-203 rounds at the dinks."

Egg, first platoon's squad leader on the island, hesitated. "That's risky. We don't know the dinks' actual location." The NCOs from second and third platoon agreed.

The other grenadier also thought it was too risky. "If they are in that cave, in order to get a good shot, we need to be on the very tip of the island. That would put us between the rest of the company and the dinks. We'll have both sides shooting at us. I'm not going," he said bluntly.

However, the discussion continued between the NCOs on the island. They finally decided that the risk was worth taking.

"Anyone want to take the other thumper?" Egg asked.

A moment of silence followed as each man weighed his options.

Finally, I volunteered first to take the reluctant grenadier's M-203. I extended my left hand, holding the shotgun chest high. He looked at me and softly said, "Good luck." Then he handed me his M-203. I briefly inspected the strange weapon. It felt awkward and cumbersome. The gunner quickly gave me an abbreviated lecture on the operation of the M-203. As soon as I thought I could operate it adeptly, Egg, the first gunner, two grunts with M-16s, and I worked our way to the tip of the island.

By this time, I was experiencing a sense of exhilaration from the adrenaline rush that combat produces. Strange as it might seem, there is a natural high during combat. In one sense, battle is the ultimate competitive sport. That is not to say I was not afraid. I was. This sport was measured not in wins and losses, but in deaths and injuries. Most of the time in Vietnam, I was either bored or terrified. Now, fueled by adrenaline, my infantry training kicked in and I did what needed to be done.

After the action, the adrenaline rush dissipated and I recalled what happened, I thought *what did I do? A guy could get killed doing something that stupid.* At that moment, the real terror overwhelmed me. *Next time I'll be more careful.* But...when next time came I was a grunt, and once more I did my job...just like all the other grunts in Charlie Company. I was no braver. On this occasion, I just took my turn to do what had to be done. If moving up to the tip of the island truly had been optional, the NCOs would never have authorized it. Our brothers in the platoon were fighting even as we debated a course of action. Supporting them was imperative. Our job therefore was to close with the enemy and kill him. The quickest way to accomplish that was to advance to the tip of the island.

We fired a couple of 40mm rounds at the bunker. Neither fell inside it. Before we could adjust fire, we were ordered to pull back to the LZ perimeter. The CO had called in Cobras from the 4th Battalion (Airmobile) of the 77th Rocket Artillery.[133] Their only avenue of approach was up the river and across the island. Our forward position risked taking friendly fire from the Cobras. In an effort to achieve superior firepower, Captain Turk had ordered us to abandon the LZ and to reinforce the embattled squads across the river from the cave. Disappointed that we did not have an opportunity to inflict any damage, we fell back to the detail's perimeter. Quickly the NCOs made dispositions to cross the river and rejoin the company.

---

133. Rutledge, "Screaming Eagle Vietnam Diary," *Rendezvous with Destiny* (Winter/Spring 1971), 35.

We rapidly waded through the shallow water between the island and the mainland. As we crossed the river, we met the RIF from second platoon returning from their long patrol. The arrival of the patrol and the cutting party enabled Captain Turk to establish a stable defensive position. First platoon's line paralleled the river. The squad returning from the island was on the left. Most of the other squads were pinned down directly across the river from the cave. Russ and his M-60 anchored the right flank. Second platoon formed a skirmish line facing up the mountain to our rear. They were spread thin since they had the longest stretch of ground to cover. However, no enemy force yet had appeared in front of them. Third platoon extended first platoon's left flank. They faced the abandoned island. The rocky terrain disrupted any attempt at an orderly front. Instead, grunts seeking cover took advantage of whatever boulders and trees they could find. This created a somewhat jumbled and untidy perimeter.

In the cutting party's haste to rejoin the company, I failed to reclaim my shotgun. I still had the M-203. Like everyone else, I sought the best cover available. My squad crammed into a tight front between third platoon and first platoon's original line. The space was inadequate for us to place everyone on the frontline. Consequently, several of us ended up in isolated spots behind the rest of the squad. I dropped into a small depression adjacent to a large tree. The spot turned out to be a poor one. I could not get a clear shot with the grenade launcher. Nor could I shoot the M-16 without firing into the rest of my platoon. The grunt with my shotgun was equally frustrated. He was adjacent to the water, but the cave on the far side of the river was out of range. He complained that when he fired, small geysers erupted about mid-stream. His grievance ended my desire to carry a shotgun. When a resupply bird dumped ammo later that afternoon, it also returned my M-16 to me. One of our wounded soldiers carried the shotgun back to Evans when he was medevac'd later that evening.

I lay on the ground looking for an opportunity to fire and wanting my steel pot. "Steel pot" was a grunt's vernacular for his helmet. Normally while humping in the jungle I suspended my helmet on the back of my rucksack and wore a soft boonie cap. I had dropped my ruck with my helmet still attached when I left to cut the LZ. Now it lay down by the riverbank in "no man's land."

Unable to participate effectively in the firefight, my mind wandered to thoughts about my helmet. Suddenly I felt totally undressed without it. The whacky idea that I might get into trouble for being out of uniform distressed me more than the danger not having it on my head posed. Occasionally I could see NVA soldiers pop up and fire. AK-47 rounds whistled over my head or ricocheted off nearby rocks.

The Cobras proved ineffective against the bunker in the cave. The topography prevented them from firing directly into the cave opening. The cave provided the NVA with a superior natural defensive position. On the other hand, our only defensive positions were tree trunks and a few large rocks. A new strategy clearly was necessary. The CO sent four of us carrying M-203s with a squad from second platoon to scale the steep ridge behind us in order to find a direct shot into the cave opening. We found a lofty position with a clear view down into the cave opening. It was beyond the place first platoon was pinned down so we did not have to fire over their heads. We deployed in a single line facing the cave across the river.

The guys from second platoon brought two M72 LAWs with them. The M72 was a 66mm high-explosive rocket in a disposable launching tube. The fiberglass tube was about two feet long in its transport configuration. To fire the rocket, one extended the tube's length by pulling out another tube inside this outer cylinder. This increased the total length to approximately three feet overall. Pressing a rectangular button on top of the outer tube fired the weapon. The tube could be destroyed after firing by smashing it against a rock or tree.

LAW stood for Light Antitank Weapon, but the LAW was effective against targets other than tanks. In Vietnam, it primarily was used against bunkers and other fixed defensive positions. The two LAWs made direct hits inside the bunker. We fired a dozen or more 40mm grenade rounds into and around the opening. The rest of the detail opened with their M-16s. The two Cobras made a final pass and peppered the jungle above the cave. The massive firepower ended the firefight. A party of grunts crossed the river to investigate. They later reported nine dead enemy bodies around the cave. One of the guys in the party told me that inside the cave was such a gory mess that it was impossible to determine the number of dead inside. One grunt rigged a mechanical ambush at the entrance to the cave using a frag with a smoke grenade fuse. He replaced the grenade pin with a sewing needle attached to a trip wire.[134]

Our casualties were surprisingly low. Charlie Company had engaged and defeated an NVA unit of at least equal size. During the initial minutes of the battle, the NVA definitely enjoyed numerical superiority but failed to capitalize on their advantage. They could have inflicted serious—perhaps fatal—damage if they pressed across the river while the company was dispersed in three separate locations. Instead, our presence surprised them. For some strange reason, the NVA did not hear the noise from us cutting the LZ. When they encountered first platoon on the riverbank, they immediately went into a defensive mode, allowing Charlie Company to reassemble. Thereafter we established superior firepower and only suffered a few wounded. Lieutenant Sweetland from second platoon and Lieutenant Smith from first platoon were hit. Smith's wound ended a promising career

---

134. Russ Kagy, e-mail message to author, February 20, 2018.

as a professional baseball pitcher. The San Francisco Giants previously drafted him. Joe Quirtess from my platoon also received a leg wound that sent him home.

The wounded were medevac'd using a jungle penetrator. The penetrator was a folding seat attached to a hoist in the door of the UH-1D/H Dust-Off helicopter. It was lowered through the jungle canopy while the helicopter hovered. The wounded were strapped to the unfolded penetrator. A cable then raised the penetrator up to the helicopter for evacuation. The procedure made the hovering helicopter extremely vulnerable to ground fire. Everyone was relieved when the last casualty was inside the bird and the pilot sped away to safety.

By the time Charlie Company terminated the action, it was getting dark. No one slept that night. We set up an elongated defensive perimeter for the night on the battlefield. The rocky ground made digging foxholes impossible. We grunts worked guard duty schedules and then hunkered down in whatever defensive shelter we could find. Captain Turk stipulated 50% alert all night. However, the extended firefight left most of the company hyped up. Instead of sleeping, the guys quietly exchanged tales of the battle and waited impatiently for morning. *Tick, tick, tick.* Every second seemed to take an hour to tick away. *Tick, tick, tick. Will dawn ever get here?* At last a faint red glow began to disclose the outline of the ridge to our east. With daylight, Charlie Company regained its swagger. Realization of its accomplishment in yesterday's firefight instilled new confidence in our war fighting abilities.

The NVA returned unheard during the night and removed the bodies of their KIA. The river's boisterous whitewater covered any noise they made and the darkness hid their movement. They also confiscated the mechanical ambush. Allowing the enemy to get that close undetected was disconcerting. We would debate what might have happened were it not for the river. Without the roaring white water, would we have spotted their presence and engaged them in battle again? Or, would they have reached our NDP unnoticed and slaughtered us? One fact was indisputable. The NVA were excellent soldiers who used the terrain to their advantage. If we were to survive, we must be equally as good or better.

Since the company had abandoned the island LZ during the engagement, there was a substantial possibility that the NVA may have booby-trapped it during the night. Captain Turk decided using it was too risky. Therefore, the company crossed the river and hiked up the mountainside. The company followed a trail used by the retreating NVA. The climb was strenuous, but not too slow. A fallen tree trunk rested across the trail at one spot, forcing the grunts to duck under it. As Russ bent down, he discovered a bloody human scalp with black hair stuck to the bottom of the log.[135] His discovery confirmed that we had inflicted real damage on the enemy we had engaged. Many grunts began to wonder why we were being extracted when we were just beginning to accomplish our mission in the Falls AO.

Charlie Company had reached the summit by early afternoon. We quickly cut a new LZ in the elephant grass on the top of the mountain. At the time, no one thought much about the view to our west. A broad green valley with more high rugged mountains on the far side was clearly visible. The nearer range was covered in dark green jungle. Subsequent ranges varied in color from bluish green to pale gray. From our vantage point, Charlie Company was looking down into the valley of death — the dreaded A Shau Valley. We were too exhausted to care. Instead, our gaze was to the east where inbound Hueys had identified our smoke and were dropping down to take us out of the Falls AO and away from the A Shau.

Little did we realize at the time that our next CA would drop

---

135. Russ Kagy, email to the author, February 20, 2018.

us into an AO equally as deadly as the A Shau. Plans were being finalized for an ARVN incursion into Laos. American ground forces would secure the route to the Laotian border.[136] Operational planners had requested the Rakkasans for a role in the campaign. Unknown to us, we would be going north into even more treacherous enemy held territory.

On January 20, Charlie Company made it back to Camp Evans for a stand-down. The 1st Battalion of the 501st Infantry took over operations in the Falls AO. The Rakkasans in Charlie Company anticipated hot showers, hot chow, and cold drinks. Most mistakenly supposed the worst was behind us. Vietnamization and the American withdrawal dominated news reports and letters from home. Army rumors speculated on when the 101st Airborne would return home.

My closest buddies in first platoon were Larry Klag from the Bronx in New York City, Joe Sej from Chicago, Illinois, and Cecil Beach from Tulsa, Oklahoma. Larry was witty and skeptical. Sharp critiques flowed from his soft shrill voice. He always seemed to stand out visually in a crowd. He constantly wore dark sunglasses when no one else did. He tied a colorful woven souvenir cloth strip around his boonie cap. Moreover, Larry loathed the army. About this time he tagged me with the nickname "Zero" after the character in Mort Walker's comic strip Beetle Bailey©. Zero was a buck-toothed private who was an honest and unsophisticated farm boy. The naïve soldier took idioms literally and misunderstood nearly everything. Larry imparted the moniker on me because of my Southern accent, my willingness to obey most commands, and because I held some simple ideas about honor and duty to country, which I still retain today. Joe Sej was a Polish kid from Chicago. His olive complexion was crowned

---

136. After the Cambodian invasion, Congress passed a law prohibiting American ground forces from entering Laos.

with a curly brown hair. He wore a perpetual grin and erupted into an exaggerated laugh when tickled. Cecil had a diminutive frame, but his cocky demeanor compensated for his small size. His square jaw and straight blonde hair were prominent facial features.

The January stand-down was markedly different from previous stand-downs. There was a lot less free time. Instead of the usual leisure, the stand-down resembled an AIT refresher course. A majority of time every day was devoted to training exercises. Most of these exercises were conducted on a squad or platoon level. Topics covered the proper care and use of the various weapons and equipment, mechanical ambushes, and early warning devices, squad ambush tactics, and radio protocol. Doc even reviewed the two types of malaria tablets and other actions necessary for good health in the bush.

We also received replacements. Lieutenant Smith's wounding left first platoon without a LT (pronounced *el tee*, slang for Lieutenant). First Lieutenant Harris, an ROTC graduate of Auburn University, was assigned the position. Harris was a great guy, but was poorly suited for combat leadership. Most of the platoon would come to evaluate him as incompetent. Second platoon also received a new platoon leader, First Lieutenant Stephen R. Metcalf. Metcalf was a 1969 graduate of the United States Military Academy at West Point.[137] He brought a gun ho attitude that most grunts considered dangerous. In reality, Metcalf already was a combat veteran who realized maintaining an aggressive edge was the best way to stay alive.[138] In time, disgruntled soldiers would make fragging threats against both. The term fragging referred to tossing a hand grenade at an individual

---

137. *Register of Graduates and Former Cadets of the US Military Academy* # 28730-1969.

138. Steve Metcalf, telephone conversation, April 21, 2017.

or an area where the victim frequently hung out. Although it gained more notoriety in Vietnam, it was equally common in earlier wars.

One of the more anticipated events during stand-down was the trip to Eagle Beach. Eagle Beach was the 101st Airborne Division's recreational area. It was located ten kilometers northeast of Hue on the South China Sea. Field units on stand-down routinely spent at least one day and one night at Eagle Beach. Activities available to the grunts included basketball games, drinking alcoholic beverages at the enlisted men's or officers' clubs, live USO shows, miniature golf, relaxing in a massage parlor, shopping at the PX, and swimming in the ocean.

We boarded a CH-47 Chinook—popularly called "s***hook"—and flew to Eagle Beach. The Chinook was a twin rotary blade helicopter that held 33 passengers and flew at a speed of 160 miles per hour.[139]

Before frolicking in the surf, I had to buy a swimming suit. I went into the PX and purchased an extra-large suit. It obviously was intended for the smaller locals. On me it had the appearance of a speedo bikini.

That evening a band called the Soul Aces played. This Philippine group played American contemporary pop music. Although they sang the lyrics in English, they obviously did not speak the language and they did not play very well. Nevertheless, their performance was a welcome distraction that everyone attended.

After the show, the grunts carried out an unauthorized unit tradition. By then nighttime covered most of the area in darkness. New officers were cunningly waylaid and lugged down to the beach. One by one, they were tossed unceremoniously into the surf. The tacit approval of more experienced officers, including the company commander, prevented serious retribution by the soaked cherry officer corps. This

---

139. Dustan, *Vietnam Choppers*, p. 75.

unauthorized baptism contributed to unit esprit de corps and fostered an accord critical for survival in the bush. The next morning the company returned to Camp Evans.

I had been promoted in December to E-4, Specialist Fourth Class. The Department of Defense gave a pay raise in 1971. The promotion and the pay raise increased my salary to $249.90 per month, plus $65.00 Hostile Fire pay. The army provided everything I truly needed to survive. My only expenses were the company beer and soda fund, beverages I bought on stand-down, and purchases from the PX.[140] Therefore, I only received $10.00 each payday. The other $304.90 was deposited directly into my savings account at Security National Bank in Cobb Center, Smyrna, Georgia. The money would give Pamela and me some financial security so we could get married when I returned home.

---

140. I bought Pamela's Noritake Ivory China set from the PX at Camp Evans. The set was selected from a catalog and shipped directly from Japan to the United States duty free. The set was the only major purchase I made in Vietnam. I also bought a Kodak Instamatic Camera from the PX at Camp Radcliff and a swimsuit from the PX at Eagle Beach.

DEWEY CANYON II/ LAM SON 719 (PART 1)

February 1971: The Mission Finally Has a Real Objective

DEWEY CANYON II[141]

The stand-down following the Falls Operation was extended an extra day. Everyone was relieved at first to have a brief reprieve from going out into the bush. By then the entire battalion had assembled at Evans. Large numbers of grunts packed the battalion area to capacity. The presence of so many troops created an unexpressed anxiety. *Something big surely is up. What could it be?* Still no one spoke of it aloud.

---

141. Nolan's book *Into Laos* is an excellent work on the Dewey Canyon II/Lam Son 719 offensive of 1971. However, his account does not always agree with my letters. The differences may be due to the fact that his primary sources did not include interviews with a member of Company C, 3rd of 187ththh Inf., 101st ABN Div. As his book points out, morale and standard operating procedures varied between different units. Unless noted otherwise this account prefers my correspondence in 1971 to Nolan as being more applicable to my experience in Charlie Company.

An American USO folk group entertained us in the battalion area. They were much better than the Soul Aces. Both groups sang "My Girl," the Motown hit by the Temptations.

I've got sunshine on a cloudy day
When it's cold outside I've got the month of May
I guess you'd say
What can make me feel this way?
My girl (my girl, my girl)
Talkin' 'bout my girl (my girl)[142]

Back in the States, the song had hit number one on the pop charts in early 1965. In Vietnam "My Girl" remained popular for the rest of the war. Whenever a band performed the hit song, troops sang along with passion. For many of us it was a pleasing reminder of the girl waiting at home. For me the lyrics conjured up emotional visions of Pamela and memories of our times together before Vietnam. Man, did it make me homesick.

Unquestionably, one song best expressed the soul of the American troops in Vietnam in early 1971, "We Gotta Get Out of This Place" by the Animals. Every band played it, accompanied by the rousing cheers of their audience. Every voice in the audience joined in enthusiastically as the band sang the chorus. Under the guise of singing a popular hit song, they verbalized their deepest and most sincere desire.

We gotta get out of this place
If it's the last thing we ever do

---

142. Written by William Robinson Jr. and Ronald A. White, Lyrics © EMI Music Publishing, Universal Music Publishing Group, Sony/ATV Music Publishing LLC.

THE FLAG WAS STILL THERE

We just gotta get out of this place
There's a better life, there's a better world
There's a better way
Don't you know, don't you know, don't you know[143]

Everyone knew that the United States no longer intended to win the war in Vietnam. And no one wanted to die for a meaningless cause. We just wanted to get out of Vietnam alive!

Then stand-down was extended a second day. Two extensions coupled with an increased training schedule and the assembly of the entire battalion made everyone apprehensive. Something was up and whatever it might be, it likely did not bode well for us grunts. Intuitively anticipating increased combat, the grunts became more restless with each extension. The presence of the entire battalion added to the tension. The men were accustomed to one or two companies at a time being on stand-down. However, during the course of this stand-down, one by one the battalion's other companies had joined Charlie Company at Camp Evans. *Something big was up.*

The protracted time in the rear produced problems within the company. In addition to its infantrymen, an infantry battalion includes numerous soldiers with logistical jobs such as clerks, cooks, mechanics, and truckers. Army dogma during the Vietnam War alleged ten men were required in the rear to support one man at the front. Most of these support troops served in specialized units, but every infantry battalion had non-combat troops, primarily in the Headquarters Company, but a few were dispersed in line companies also. Conflict between the grunts and the non-combat members of the battalion flared up during this stand-down.

---

143. Written by Barry Mann and Cynthia Weil, Lyrics © Sony/ATV Music Publishing LLC.

The grunts derisively referred to these fortunate troops as REMFs (pronounced *reamfts*), a shortened form of Rear Echelon Mother F***ers. The grunts' disdain no doubt was born from a combination of jealousy and superiority. Grunts were jealous of the REMF's relative safety in the rear and their comparative luxury. Grunts assumed a REMF wore clean uniforms, took daily showers, ate three hot meals a day, drank cold beer and soda, slept in dry beds, employed hooch maids, enjoyed regular sex with Vietnamese girlfriends or prostitutes, and listened to their favorite music on AFVN. While the grunt's perception was amply biased, life in the rear and life in the bush were incomparable. Grunts rarely changed clothes more than once a month, showered only when it rained or were near a river, ate cold C-Rations, drank iodine tasting water, slept on the wet ground, reminisced girls from faded photographs in their wallets, and listened to the symphony of exotic birds and other indigenous wildlife. Grunts hated most REMFs with a vengeance.

Furthermore, grunts considered themselves infinitely superior to REMFs because a grunt had been "in the s***" and survived. The slang expression "in the s***" covered a host of meanings too vast to enumerate. Everything from the horror of seeing comrades killed or wounded to the misery of weather was included in this simple profanity. The grunt had endured more adversity than a REMF was capable of imagining. How can inconvenience caused by a broken generator compare to the agony of climbing a steep mountainside in the monsoon rain with a 70 pound rucksack on your back? Or how can the guard duty on the green line at Evans compare to the vulnerability on ambush in the boonies? So grunts gloried in the CIB on their uniform and generally antagonized a REMF whenever possible.

Racial friction also resurfaced. The battalion had experienced a serious incident the previous year. African-Americans had organized into a militant "black power" society. The hostility came to a head in

December when Colonel Sutton broke up the society's power base. A number of the militants were arrested, but some white officers in the senior command structure were transferred out of the battalion as well. Now race tension reared up once more. It was never a problem in the bush. Race differences were non-existent in the close proximity of NVA soldiers. The NVA wanted to defeat the Americans. Doing so compelled them to kill American soldiers. Therefore, the color of a man's skin or his ethnic background did not matter to the NVA. They wanted to kill all Americans. Nor did it matter to an American grunt. He simply wanted not to be killed. Therefore all that counted to a grunt was how well a man performed his job. Survival required all the grunts in a unit to work together. They quickly learned to trust and depend upon each other. Without absolute trust and assurance in each other, an infantry platoon lost its unit cohesion and combat proficiency. When it did, people died. In the field men were American grunts, not black or white grunts, not Hispanic or French Canadian grunts…American grunts. However, in the rear that unity sometimes broke down as grunts met the ethnic peer pressure of REMFs' prejudice and contact with the world outside one's platoon or company. African American grunts transformed into conceited "Soul Brothers" and rural whites converted into condescending redneck hillbillies.

Drugs and alcohol contributed to the general unease. Field unity was lost as the company divided into "heads" (drug users) and "juicers" (alcohol drinkers). The use of either was taboo in the bush. Grunts prided themselves in their ability to cope with the misery of the bush. The use of stimulants or depressants abrogated this bizarre pride. More importantly, however, no one was willing to risk his buddies' wellbeing since both impeded mental and physical performance. But within the relative security of Camp Evans, a majority of grunts turned to one or the other to release the stress built up by weeks in the boonies.

Added to the general unrest were the innate rivalries that exist between units. Infantry units hold non-combat units in particular contempt. A number of fights erupted. On one occasion, a number of grunts sought admittance to a transportation unit's enlisted men's club. When told that only unit members were admitted inside the premises, the grunts forced their way inside, beat up several patrons, and dismantled the facility. Fortunately, there were no serious injuries. Although the incident involved a large number of grunts, none of the attackers could be identified. So no one was punished for the incident.

Charlie Company's First Sergeant called a company formation late in the afternoon of January 29, 1971. Such stateside military routines were extremely rare in Vietnam. Most combat units considered them trivial and unnecessary. The order to assemble at 1800 hours (6:00 p.m.) generated loud grumbling and open hostility.

"What's this chickens***?"

"I ain't going to no f***ing formation!"

"This is f***ing chickens***!"

Draftees composed a large number of the men in the ranks of the army in Vietnam by 1971. For the most part, these men resented being in the army. They pursued any opportunity to flaunt their antipathy. Military protocol often was slack at times. These men knew their job in combat and did it well. Their survival required them to do so. Nonetheless, they were civilians fighting in a war their country already turned against. Consequently, their sole motivation to be soldiers was survival. The men in Charlie Company realized that in the bush being a good soldier was essential to staying alive. Grunts inevitably encountered NVA in the bush. The NVA were superb soldiers. Most had far more combat experience than their American counterpart did. The NVA soldier wanted to go home as much as an American grunt did but the only way an NVA soldier could go home was to kill all the American grunts in the bush. Thus, if an American grunt wanted to make it home, he had to be a better soldier than his enemy, the NVA soldier.

The company First Sergeant was an old "brown boot" paratrooper named Dayton Herrington. He enlisted in the army in 1948 and fought in Korea with the 180th Infantry Regiment. He had introduced himself to Charlie Company with a memorable speech.

"Gentlemen, I love the 101st Airborne Division. I love the United States Army. I love my wife. And gentlemen, I love them in that order."

When Herrington retired in 1978, he had served 16 and half years with the 101st Airborne Division.[144] The slang designation for a company first sergeant in the army was top sergeant. Hence, soldiers commonly addressed the first sergeant as "Top." Herrington did not like the term.

"Gentlemen, you will address me as 'First Sergeant,' not 'Top.' A top is a child's toy and I assure you, I am no child's toy."

At the time, the reaction from soldiers in the ranks was less than favorable. Most immediately concluded he was a gung-ho lifer who would impose senseless spit-and-polish procedures on the company. In time he earned the esteem of most grunts. Instead of a martinet, he maintained discipline by example and focusing on issues that improved the combat efficiency of the company. He was nominated and won the Soldier of the Year competition for the division that year.[145]

---

144. Career facts on Herrington taken from Yvette Smith, "Retired Sergeant Major Soldier for life," (October 18, 2013) Official U.S. Army website (http://www.army.mil/article/113451/Retired_Sergeant_Major_Soldier_for_life/). Accessed July 31, 2017.

145. *Ibid.* In time as more and more units returned stateside, even discipline and morale in the elite 101st ABN Div crumbled. Too many replacements came from

A mutual respect for each other gradually developed between this career soldier and the draftees in the ranks.[146]

Charlie Company formed up at the appointed hour. Two ranks of irritated grunts stood at ease on the company street in front of the company orderly room. First Sergeant Herrington stepped out of the building and shouted, "Company! Attention!" The formation begrudgingly snapped to attention as the Company CO followed Herrington outside. The CO nodded at Herrington. The First Sergeant bellowed, "At ease!"

A long silence trailed the command's execution. The grunts in the ranks shuffled nervously as we awaited the purpose for this extraordinary assembly. Herrington finally spoke in an audible crisp voice.

"God d*** it! I just got my a** chewed out on count of you f***heads. The CO received a d*** report that some f***ing transportation troopers got the s*** beaten out of them yesterday! Their f***ing company commander complained to the battalion commander about some d*** black eyes and f***ing bruises. Claimed it was Rakkasans. Said you destroyed his f***ing men's d*** club. H***, he has to have it right! You are the only sons of b****es dumb enough and mean enough to do this kinda s***. Every d*** last one of you ought to be f***ing article fifteened!

"D*** it! Last night two of you s***heads beat the s*** out of each other. You d*** idiots are such f***ing d*** fighters that if you don't

---

units being withdrawn, units where discipline already had broken down. These men did not assimilate into the division esprit de corps. In late 1971, just before the 101st returned to Fort Campbell, Kentucky, Herrington was jumped and beaten up by unknown soldiers.

146. Alan Davies, Russ Kagy, and myself reunited with CSM Herrington at the 101st Airborne Division Association 2018 Snowbird Reunion, in Tampa, Florida, February 8-10, 2018. He proudly introduced us to others as "his soldiers."

have a f***ing enemy to fight, then you must f***ing fight each other. Therefore the battalion commander has determined it obligatory for you to return to the field. The entire f***ing battalion will board f***ing choppers before dawn for a new combat operation."

The CO spoke briefly. His speech included the usual pep rally style indoctrination about how proud he was to command such fine troops and how he knew we would accomplish our mission.

Then Herrington notified us that the Military Police had sealed off our battalion area for operational security. No one would be allowed to enter or leave the battalion area. The news spawned quick glances towards roads and paths visible from our formation. We could see Military Police vehicles indeed blocked all of these. Rows of fresh concertina wire were stretched across every opening in the wire surrounding the battalion area. We assumed they also blocked the exits we could not see. Then Herrington announced that everyone in the company was confined to barracks when the formation was dismissed. He concluded by strongly encouraging the entire company to attend a battalion briefing at 1900 hours (7:00 p.m.). For all NCOs and RTOs the meeting was mandatory. *Something big* obviously was up. The irregular stand-down, the uncharacteristic security, and an atypical invitation to a battalion briefing sparked my curiosity. *Something big, really big,* was up. I showed up for the briefing.

I found a spot and sat down on the floor of the battalion briefing room, I listened intently as Colonel Sutton revealed our battalion's role in an operation called Dewey Canyon II. It was the first phase of a larger operation known as Lam Son 719. In Lam Son 719 the South Vietnamese Army would cross the Laotian border and sever the Hô Chí Minh Trail. On D-Day of Dewey Canyon II, American forces from the 101st Airborne, the 23rd Americal Division, and the 5th Mechanized Infantry Division would attack into the Khe Sanh Plateau, securing QL 9 all the way to the Laotian border. American forces also would reopen the old Marine firebase at Khe Sanh. It would serve as a logistic operational base for the invasion into Laos.

An endless roll of NVA regiments was called and reported as operating in the area into which we were going. "Army intelligence places the NVA 2nd Division, the 304th Division, the 308th Division, the 320th Division, and the 325th Division in the AO...etc."

D***! This is some deep s***.

A total of 22,000 NVA troops were estimated in the region of Dewey Canyon II and Lam Son 719 operations. Thirteen thousand of these were combat soldiers in the main line units identified during the briefing.[147]

The noise faded. People stopped talking. The normal shuffling of feet and arms ended. An eerie silence gripped the room. Everyone now strained to hear what the battalion commander was saying. This was not a routine CA. The entire battalion, including Charlie Company, was going into Indian country, *very bad Indian country, d*** bad Indian country.*

Indian country was slang for areas controlled by the North Vietnamese. The expression likely came out of the countless Western movies our generation grew up watching in theaters and on television. The metaphor evoked the memory of the Indian victory at the Little Big Horn in 1876. In that battle, 268 U. S. Army soldiers under Lieutenant Colonel George Armstrong Custer died fighting thousands of Sioux, Cheyenne, and other Native American warriors. The intelligence report unhesitatingly conjured up a gloomy comparison with Custer. We all clearly saw that where they were sending us now surely had the potential to be as bad as the Little Big Horn. If the estimated strength

---

147. 101st Airborne Division (Airmobile), Final Report; Airmobile Operations in Support of Operation LAM SON 719, 8 February-6 April 1971, Volume II (1 May 1971), I-4.

of the NVA were correct, we too would face overwhelming odds.[148] More than one man present in that briefing imagined we might suffer the same fate as the massacred cavalrymen 95 years earlier. Moreover, from our first day in the battalion, the Rakkasans' role on Hamburger Hill had been drilled into the very fiber of our being. Almost two years had passed since that battle—and none of us were in the battalion then—but the specter of its carnage hung over us like the Sword of Damocles. Charlie Company suffered 80% casualties in that ten-day fight. Now the 3/187 Infantry once more was heading into bad, bad Indian country…into the same area where the Marines had been surrounded and endured a 77 day long siege in 1968.

But where exactly? Sutton already had summarized Dewey Canyon II and Lam Son 719, but his remarks had not registered fully yet. The information on enemy strength overshadowed everything else. An orderly pulled down a large roll-up map. *S***!* The letters D M Z dominated my gaze. Sutton stood in front of the map with a pointer. He meticulously divulged plans about the South Vietnamese invasion of Laos. *This is bad. What about us?* Finally, Sutton covered the battalion's role in Dewey Canyon II. Our mission involved securing vital bridges along QL 9 near Firebase Shepherd. QL 9 was the main east-west highway between the coast of Vietnam and the border of Laos. Along the way it passed the combat base at Khe Sanh and the area known as the Rockpile. The mention of these two places recalled memories of the bloody battles the Marines fought in those locations during 1968. *This is bad, real bad.* Shepherd was only four miles southeast of Khe Sanh.[149] According to the briefing we were headed into a

---

148. The NVA was able to field superior numbers. However, the majority of its forces deployed in Laos against the ARVN invasion, not against the American forces in South Vietnam.
149. Charles McClintock, "Dewey Canyon II/Lam Son 719," *Rendezvous with*

region the enemy had more or less controlled for three years and we were going to face a foe that greatly outnumbered us and conceivably was better equipped than us. *D*** this is real bad. We're getting in some deep s*** this time.* We only had a few hours to prepare before *stepping into this deep s****.

During the initial phase of Dewey Canyon II, operational control of the Rakkasans passed temporarily to the 1st Brigade of the 5th Infantry Division (Mechanized). The 4th Battalion of the 3rd Infantry, 23rd Infantry Division likewise was put under the operational control of the 1st Brigade of the 5th Infantry Division. The mission assigned to the 5th Mech's First Brigade was reopening QL-9 from Quang Tri to Khe Sanh by seizing key terrain between Ca Lu and Khe Sanh.[150] Once Khe Sanh was secure, the enlarged brigade was to take control of the highway from Khe Sanh to the Laotian border. This assignment needed to be accomplished in only a few days.

Everyone was quiet as we walked back to the company area. We could see MP (Military Police) roadblocks set up at every entrance to the battalion area. The atmosphere was somber in the platoon hooch. Sergeants conveyed the news to those who missed the briefing. No exaggeration was necessary. The name Khe Sanh was sufficient. A veil of foreboding descended over the platoon. All discord and foolishness vanished. Our fate was now inescapable. All anyone could do was prepare for its arrival.

The grunts began to check their gear. They emptied their rucksacks and inventoried its contents. Supply issued additional ammunition. M-16 rounds came in a seven-pocket cotton bandolier. Each pocket contained a cardboard sleeve that held two 10-round clips of

---

Destiny (Winter/Spring 1971), 29.

150. Recommendation for the Award of the Presidential Unit Citation for 1st Infantry Brigade, 5th Infantry Division (Mech) 18 June 1971, 1.

5.56mm bullets. The grunts removed the clips and inserted the bullets into 20-round metal magazines. They placed the full magazine back into the bandolier rather than insert it into an issue magazine pouch. The bandolier held more ammunition, was lighter, and less cumbersome than the pouch. Consequently the individual soldier could carry more ammo using bandoliers than regulation gear.

That night most infantrymen packed two or three bandoliers of 20-round magazines. Everyone also lugged one 100-round belt of 7.62mm rounds for their squad's M60 machine gun. In addition, each man took five or six M26 fragmentation grenades and one colored smoke grenade. They removed the grenades from the cardboard canisters in which they were packed. Most grunts suspended two to four of the grenades on their webbing for quick access in a firefight. They carried the rest inside one of the rucksack pockets. Grenadiers carried M-203 grenade ammunition in a nylon vest that held 20 40mm rounds. Everyone stuffed one Claymore mine inside his rucksack. Some riflemen also packed an M72 LAW on the top of their ruck.

C–Rations were distributed. Each man received one and a half cases. He then bartered with comrades to get the meals he liked most. The 18 meals were double what they customarily humped, but the scope and nature of the campaign stretched logistics to their utmost. Time between resupply birds would be longer. Officers allowed their men out of the barracks long enough to fill their canteens from a nearby water buffalo. Grunts lined up to use the nearest piss tube or outhouse. Supply clerks attempted to replace all missing equipment. This preparation commotion consumed the evening. A few guys stole a few winks of sleep, but most were too busy to even stretch out on their cot. Besides, fear and apprehension deterred any real relaxation that night.

In addition to ammunition, food, and water, most soldiers carried an M-60 ammo box filled with personal items. The sturdy cans made excellent waterproof containers. Stationary, pens and pencils, an inexpensive Kodak Instamatic Camera, a razor, and other personal items could be stored safely inside. My wallet was stuck inside an old clear plastic bag in which radio batteries had been packaged. Extra socks, a poncho, and a poncho liner were crammed inside the main compartment of the rucksack. Together with his steel pot, a grunt's average load now was between 70 to 100 pounds.

I didn't attempt to sleep. D-Day was only hours away. Once my gear was squared away, I lay down but didn't close my eyes. I stared at the ceiling without observing the tin sheets or wooden rafters. Instead I saw my past and the future life I hoped to have. I recalled neglected memories of Daddy and Mother, Reita and Lisa. I mentally reconstructed an image of my dearest Pamela. *What would become of Pamela if I didn't make it? Enough of such morbid thinking! I best concentrate on my job or I most certainly would be killed.* But concentrating on my job did not guarantee my survival. So, my thoughts turned to God and I prayed silently. I didn't utter specific words as much as I just meditated upon my sovereign Lord. My destiny rested in His hands.

While most grunts did not comprehend the overall strategy, we understood one overarching reality. For us the war was about to escalate radically. Some of us indisputably would never return from this operation. Each individual handled his emotions in whatever way best suited him. One guy even was polishing M-16 rounds with Brasso®[151] before inserting them into a magazine. Most men remained silent. They sat on their cots, either working on equipment or meditating about tomorrow. A no-nonsense attitude overshadowed everyone. Occasionally someone attempted to ease the tension with a course

---

151. Brasso® was a commercial metal polish. Soldiers routinely used it to polish the brass components on dress uniforms.

joke or smart-alecky jab. Their effort yielded a muted response and then the heavy silence returned.

D-DAY

Long before daylight, a convoy of deuce-and-a-half trucks pulled into the battalion area. Their appearance only added to the sense that this mission was something different. Normally we walked to the battalion helipad. That helipad was too small to handle the number of choppers required for this operation. Therefore the grunts piled into the back of trucks for the brief ride to Camp Evans' main runway. Numbers were assigned to each grunt that corresponded to specific helicopters. We loaded into the trucks accordingly. The truck beds were higher than the floor of a Huey so loading troops with rucksacks weighing over 70 pounds was cumbersome. With few exceptions, a grunt removed his ruck and slung it up onto the truck bed. Then he put his foot in the stirrup of the lowered tailgate and reached up. A buddy in the truck grabbed his hand and pulled him up into the truck. Often another grunt on the ground simultaneously shoved him up. Even with assistance, getting up was difficult. The wooden bench inside the truck bed was too narrow to sit on with a fully loaded rucksack. Therefore, field packs were placed directly in front of their owner, who clutched its straps in one hand and secured his weapon in the other. The trucks finally were loaded and the convoy snaked its way across the base's dirt roads to the main runway.

The truck's cargo bay was uncovered, but darkness hid most sights. The scene on the tarmac contrasted starkly with the surrounding vista. Outside the wire that enclosed Camp Evans, the night was abysmally dark. Inside the wire, the radiance of an occasional lamp bulb revealed some of the base's buildings. However, most grunts caught only a fleeting glance at the distant structures. Instead, all eyes focused on the runway. It glowed with red, green, and white lights. These were too dim to illuminate their surroundings fully but there were too many to preserve darkness. A murky radiance exposed the runway's aircraft and activity. It was easy to see even in the limited light that the runway was crammed with seemingly endless rows of UH-1H Huey helicopters. The sheer number of helicopters astounded most of the grunts. Few, if any, had ever seen so many Hueys in one place. The size of the armada posed more evidence that this was not a routine mission. The convoy broke apart. The trucks drove in between the rows of choppers and emptied their human cargo.

Grunts unloaded from the trucks. Encumbered with the weight of equipment and supplies, each man required assistance getting down out of the tailgate and onto the tarmac. Once on the ground, he adjusted his load and quietly moved to an adjacent helicopter. NCOs allocated five man teams to each bird. Lieutenant Harris and his RTO, Big Steve, Uncle Al, and I headed towards the nearest Huey. As I approached my assigned bird, I was startled to see the yellow and black 1st Cav patch painted on its nose and tail. The First Air Cavalry Division was stationed in the southern part of Vietnam near Saigon and Bein Hoa, not in I Corps. *How and why was this mysterious bird up here?* A serious atmosphere prevailed on the runway, but our inquisitiveness about the enigmatic bird was too much. We exchanged whispered words with its crew as we stood around waiting for the signal to load up. The bird and its crew indeed were from the 1st Air Cav. They had flown to Evans directly from their base in Military Region 3. They arrived at Evans around 0200 hours (2:00 o'clock in the morning). They checked and refueled their helicopter upon arrival. Then they snacked on cold C-Rations and tried to catch a brief nap while waiting for the convoy of grunts.

Grunts stood humped over beneath the weight of their rucksacks as they waited to climb into their assigned seats. Some needed the aid of a buddy to scramble into the chopper. Once settled on the floor of the helicopter's cargo space, the infantrymen sat and waited for takeoff. Aircraft crews checked their aircraft one last time. Anxious truckers slowly guided their vehicles away from the area before some dink on Rocket Ridge launched a missile into the unusual activity.

We were airborne before daylight. Approximately 100 slicks carrying the four line companies of the 3/187th Infantry flew in a formation of two staggered columns. In addition, there were a dozen or more Cobra gunships plus the battalion's C&C bird. C&C was an acronym for Command and Control. Commanding officers used C&C helicopters to supervise their unit's movements. The formation headed north. Near Dong Ha it turned abruptly west. By then dawn was dispelling the night's darkness. I sat on the floor with my feet dangling out the door.

A trip into the bush always was a time for introspection. A human being had to prepare mentally for what might await him at the LZ. One could not escape the facts. The flight terminated in hostile territory, Indian Country. The intelligence information about NVA regiments in the new AO replayed repeatedly in my mind that morning. People who wanted to kill us likely waited for us at our destination. The end of my life might only be minutes away. I tried to evade thoughts of dying by contemplating other matters. Thinking about Pamela only made me wonder if I ever would see her again. It was useless. The fear lingered. Instinct touted flight but the elevation and movement of the helicopter prevented escape. No one would get out of the helicopter's cabin until it touched down at the designated LZ. I would make the CA. Whether I lived or died still was in God's hands and not mine.

The psychological contemplation mingled with visual scrutiny of the ground below and other helicopters. The optical investigation distracted from the tension and made the flight endurable. Looking straight ahead, I saw another formation of helicopters in the distance. It was approximately the same size as ours moving parallel with us. I watched it briefly. Then I strained to look behind me. I barely could glimpse a third formation over my shoulder. It also was moving parallel with ours, but I couldn't make out its size or what kind of helicopters comprised it. By then the sun had fully illuminated the panorama around me. I peered down at the ground beneath me.

We were flying west directly above QL 9. I observed a column of M-48 tanks, Sheridan tanks, APCs, and assorted other vehicles on the road below me. The column seemed to stretch as far as I could see in either direction. In the elephant grass on either side of the column, armored fighting vehicles secured the column's flanks. I had never seen anything like what I was looking at in that moment. It was the largest display of military might that I have ever witnessed. As I continued to observe it, a sense of dominance swept over me. *How could the NVA possibly resist such a force?*

As I meditated on the observable firepower, the helicopter formation began dissolving as individual companies set out toward their specific objectives. The 24 slicks carrying Charlie Company peeled off from the formation and headed for our objective.[152] As it moved away from the main body, the smaller helicopter formation spread into the familiar V formation. Suddenly uncertainty about what was ahead replaced the sense of overwhelming power. Concern about the unknown always gripped you going into an LZ. On this occasion, memory of the previous evening's briefing multiplied the trepidation. So too did the dispersal of all those helicopters that I watched a few minutes before. I looked down. The road with all those vehicles I saw

---

152. The roster for January 1971 listed the company's strength at 120 men plus officers. No roster for February or March is available.

only a short time ago had vanished. All I could see was endless hills covered in elephant grass or jungle. My fear returned with vengeance. I glanced at the stoic expressions of other grunts in my chopper. I wondered if they could see my fear. Were they as frightened as me? My attention intuitively began rehearsing my actions during the first minutes on the LZ.

The LZ was large enough to accept five helicopters at a time. Consequently, Charlie Company would land in five successive waves. Each wave would be spaced three minutes apart. First platoon would land in the first two waves. My bird was in the first landing. As the other companies headed towards their objective, the five waves in our CA slowly spaced themselves apart to accommodate the timing of their landing. My heart was pounding. *Slow down heart! We don't want anyone to hear you.* Although I knew intellectually that the others in my chopper were experiencing similar anxiety, I did not want to give away any sense that I was afraid.

The olive drab automated birds began their descent. I fixed my sight on the approaching LZ. A pair of Cobras rose into the grey sky after their last rocket run. White smoke rose from burning grass on the LZ. The day had turned murky and rainy. A gray mist detained the smoke and obscured vision as the chopper sank into the haze. Door gunners on the Hueys sprayed the foliage around the LZ. Cobra gunships circled around, looking for targets of opportunity.

Inside, my stomach was in knots. Although almost paralyzed with fright, I resigned myself to the inevitability of leaving the superficial safety of the helicopter. The only antidote to this anxiety was getting on the ground. I just wanted to "get it over with." The sooner we were on the ground, the sooner the helicopters behind us could land. A sense of security would return once the company reassembled.

As the bird descended, I slid out of the door. My feet landed square on the Huey's skid. I straightened up slightly and readied myself to jump. The Huey slowed down and hovered briefly a few feet above the blackened earth. I stepped forward. Gravity employed the heavy load of my rucksack to hurl my body down onto the ground. Somehow I managed not to stumble upon impact. Instinctively I bent down to avoid the rotary blade and then raced away from the chopper. I had to get clear before the next bird touched down. I headed directly to the edge of the LZ and my preassigned spot on the perimeter. A light rain smothered the remaining flames but smoke still lingered on the LZ. Breathing became difficult and tears formed in irritated eyes. Defensive positions were in the unburned grass. They promised improved breathing, cover, and concealment. Big Steve and Uncle Al trailed just behind me. Lieutenant Harris and his RTO rounded the nose of the bird and sprinted to the edge of the LZ. I spotted Egg running from the bird behind mine. His familiar voice shouted, "First squad! Over here!"

I ran towards Egg. He motioned to a place in the grass. I was panting as I dropped to the prone firing position. The sprint had been short but my heart was beating so hard I thought it was going to explode. I pulled back the charging handle on my M-16, released it smartly, and loaded a round into the chamber. I made a quick survey of the ground to my front. *Elephant grass. Nothing but elephant grass.* Finally feeling safe for the moment, I looked around. A couple of yards to my left, I could see Big Steve tinkering with a belt of ammo feeding his M-60. I raised my left hand and released my rucksack. Freed from its cumbersome weight, I wiggled into a more comfortable prone position. Still nothing moved in my front. The company perimeter grew stronger with each successive wave of helicopters. The last bird dropped its passengers and lifted skyward. Charlie Company was fully prepared to fight but the locality was strangely silent. For the first time that day I felt chilled by the moist air.

The anticipated NVA heavy resistance failed to materialize. In the peaceful surroundings we took time to gobble down cold C–Rations. We had not eaten since dinner the previous afternoon and were hungry. After consuming the meal, we buried the cans and trash. Then Charlie Company set out for the first day's objective. First platoon was in the lead. I took point. Larry Klag walked slack. Big Steve and Uncle Al ambled behind us. Harris and his RTO led the rest of the company. We moved in single file.

Elephant grass covered the rolling hills over which we traversed. Elephant grass generally grows to between 3 and 7 feet in height, but can reach up to 12 or 13 feet. Its razor sharp blades regularly lacerated exposed skin. Therefore, grunts always unrolled the jacket sleeves when moving through this vegetation. But the safeguard against cuts also increased body heat and added to the misery of humping. Our objective was a small hilltop approximately one and a half kilometers from our LZ. The company would establish a defensive perimeter there from which to patrol the area.

As the company moved through the grass toward the hill, I encountered a dirt path. It skirted the base of the hill staying at a constant elevation. This kept the trail level but created a parabolic curve in plan. We intersected the parabola at its vertex. This enabled us to see 10-15 yards down the path in either direction. The company halted. Larry and I crept slowly down the path to our left to investigate. Big Steve and Uncle Al set up their M-60 machine gun to cover the path to the right. They could turn it quickly to support Larry and me if necessary. They lay prone on the ground. Steve peered down the barrel of the machine gun. His left hand rested on the neck of the weapon's shoulder stock as his right index finger softly pressed against the trigger. Uncle Al was to his left, lying just off the trail. He had removed one of the belts of M-60 ammo that was wrapped around his torso and was ready to clip it to the end of the belt feeding into Steve's gun. The M-60 was nicknamed "the pig" because of its weight and because of the need to feed it more ammunition constantly when in use. Its appetite for bullets gave it exceptional firepower on the battlefield.

As point man, my primary responsibility was booby traps. So I cautiously scrutinized the ground where I was about to step. Nevertheless, with each step I quickly scanned the trail ahead of me. As slack man, Larry was only an arm's length behind me. He was responsible for protecting me from ambush or snipers. His eyes constantly perused from side to side as he peered into the vegetation around us. I took only a couple of steps on our investigation of trail and stopped. I could hardly believe what I spotted. A large cat's paw print was clearly pressed into the soft earth! It was maybe four and half inches wide and five and half inches long, but it seemed much larger. Several more were visible further down the path. I stopped and dropped to one knee. Larry instinctively did the same. I turned my head over my left shoulder and whispered softly to Larry, "I've got tiger tracks up here." Quickly my vision reverted back to the trail. However my attention now was focused on tigers rather than booby traps.

"What kina twacks did youse say?' Larry asked in his soft spoken Bronx nasal twang.

"Tiger tracks," I answered, turning back only long enough to utter the answer.

"Naw, man, becawse de-ah's no tigahs out he-ah," he retorted.

When I insisted, Big Steve joined in. Big Steve was the largest man in the platoon. His size was intimidating, but he was a man of few words. When he did speak, his voice was gentle and normally muted.

"Stop kiddin. There ain't no tigers," he said sharply. This time his voice hinted at some irritation at me for stopping our advance for what he perceived to be nonsensical teasing.

By this time Lieutenant Harris had come up to check out the

reason for the delay. He squatted just behind Steve and Uncle Al. They told him that I was imagining things. In the absence of any signs of the NVA, they were having a good laugh at my expense.

"Well if I don't have tiger tracks, I have a mighty big putty cat tracks," I snapped back.

I had no sooner spoken than about ten yards in front of me a large tiger emerged out of the grass. That cat was one of the most magnificent creatures I have ever seen in my entire life. It moved slowly and deliberately. Taunt muscles rippled with each graceful move. Vertical black stripes streaked burnt orange fur and a snow-white belly. The wild cat paused briefly to examine us. Its yellow-green eyes scanned us as if deciding whether or not we were on the menu for lunch. Having rejected us for its culinary appetite, the tiger moved on at the same tempo as previously. It quickly vanished in the tall grass.

I knelt there speechless for a moment. Larry also was frozen, thunderstruck at what just happened. Finally everyone relaxed. Steve and Al stood up. Al hadn't seen the tiger. The three of us who had witnessed the cat displayed a colossal grin. Then Larry joked about taking a trophy photograph standing over a dead tiger. We quickly decided that an M-16 would not kill the big cat. Then someone suggested the M-60 might do the job. A radio call from the company commander quickly put to an end to our momentary hunting diversion. The delay had lasted long enough. It was time for us to move out and occupy the hill. Besides, the NVA was out there somewhere. If we overlooked them, we soon would be stuffing body bags.

For the next few days everything was quiet. We spent three days searching for some NVA mortars that were harassing elements of the 5th Mech on QL 9. The mortars portability made the search an exercise in futility. Since the enemy moved whenever we approached, the best attainable result was calling in artillery strikes upon suspected mortar positions while they were firing. Although we never captured the NVA mortars or recovered the corpses of their crews, the pestering fire diminished.

* * *

About this time my parents received a letter from John Little. John had ended up in the 173rd Airborne Brigade at Bong Son. Bong Son was located on QL 1 approximately 75 kilometers northwest of Qui Nhon. In his letter he told Mother and Daddy that the 101st was being deactivated and that I would be home soon. Unaware yet of the incursion into Laos and the Rakkasans role in it, they wrote a letter full of unwarranted anticipation of our reunion. I assured them that the rumor of our deactivation was "nonsense."[153] Reports from Lam Son 719 and Dewey Canyon II dominated news broadcasts by the time they received my letter. Dreams of any forthcoming reunion quickly evaporated. Instead their apprehension for my safety amplified with each news report.

* * *

On February 1, Major General Thomas M. Tarpley took command of the 101st Airborne Division from Major General John J. Hennessey. Hennessey graduated from West Point in 1944 and served as a platoon leader in the European Theater during World War II. He was a staff officer during the Korean War. He served one previous tour in Vietnam (1965) as a brigade commander with the 1st Cavalry Division (Air Mobile). Tarpley also graduated West Point in 1944 and served as a platoon leader in Europe. In Vietnam, Tarpley previously commanded a battalion with the 5th Infantry Division (Mechanized)

---

153. Letter to my parents dated "12 Feb 71."

and a brigade with the 25th Infantry Division. The change in commanding generals went unnoticed by Charlie Company's grunts.

Over the next few days contact with the NVA forces within the 3/187th Infantry's AO increased. A post-action report on NVA tactics in Laos described the enemy's dispersal there:

An effective technique used by the NVA is employment throughout the operational area of ten-twelve man combat teams armed with small arms, at least one 12.7mm machine gun, at least one 82mm mortar, and one or two RPG rocket launchers. Positioned on or near critical terrain, located in bunkers and trenches, well-supplied with ammunition, these combat teams attack by fire aircraft and infantry operating within their weapons range. The teams are capable of placing 12.7mm machine gun and 82mm mortar fire on virtually every friendly position, landing zone, and pick up zone in the Lam Son 719 operational area.[154]

Similar dispersion of NVA units occurred east of the border as well. By the time the ARVN force crossed the Laotian border, the North Vietnamese had massed approximately 60,000 troops in Laos to oppose the invasion. This was about twice the number of South Vietnamese troops going into Laos.[155]

After several days wasted searching for the elusive mortars,

---

154. "Airmobile Operations in Support of Operation Lam Son 719 - March 20, 1971," Headquarters, 101st Airborne Division (Airmobile) in The Vietnam Center and Archives, Texas Tech (http://www.virtual.vietnam.ttu.edu/cgi-bin/starfetch.exe?KkufYUz@TOqERR@MQsQIKkJ6ctA6uhrC@Y7IpmXs82xjq.H6NqgM-jUvBjShShHBnoqPkbYhRI4BM2M86fT0hHMKl2a@b9YkqQGmJrRP.RvVzIx-1En35ovg/2960107001a.pdf). Accessed August 1, 2016.

155. Graham A. Cosmas, *MACV: The Joint Command in the Years of Withdrawal, 1968-1973*, United States Army in Vietnam (Washington, D.C.: Center of Military History United States Army, 2006), pp. 330-31.

operational commanders decided to change the company's AO. Bravo Company needed a blocking force in their hunt for some enemy 122mm rocket launchers. A flight of Huey helicopters picked up Charlie Company and moved us to the new location. Our CA was textbook. The choppers touched down briefly. Grunts jumped out. Hunching over they raced beneath spinning rotary blades. They spread out and rapidly established a secure perimeter.

As soon as the entire company was on the ground, we moved toward an assigned location to support Bravo Company. The new AO virtually was identical to that where had been before the transfer, more tall elephant grass. The men rolled down their sleeves to protect their skin from the grass's razor cuts. Nevertheless, travel was easy and we made brisk progress. Later that day Bravo reported the capture of the rocket launchers. The next morning Charlie Company returned to searching for NVA. Instead of contact with hostile troops the company continued wandering through an endless ocean of elephant grass and bamboo.

As the ARVN incursion into Laos opened, combat intensified for American units involved in Dewey Canyon II. Word passed through the ranks that the ground commander had designated Charlie Company as a reaction force for all American units participating in Dewey Canyon II. Then the company CP received word that an American unit was in heavy contact. One NVA unit had stopped its evasion and was fighting back. The CO's voice rang out with an all too familiar call, "Charlie Company! Saddle up! Birds inbound." Then one of the senior NCOs interjected, "Boys, this one going to be hot! Some f***ing Americal a**holes have stepped in some deep s***." On February 17 we loaded up for our third CA in less than three weeks.

PURPLE HEART HILL

The sun already was low on the western horizon when the choppers began extracting Charlie Company. There were insufficient helicopters available to transport the entire company to its new AO in one trip. Too many American choppers were committed to the ARVNs in Laos. Half the company would have to wait on the PZ until the available helicopters could drop off the other half of the company and then return to pick them up. The CO assigned first platoon to the initial air assault. By the time that we reached the LZ, sunset diminished visibility dramatically. Below us, the lush green landscape faded into dreary charcoal shadows. The choppers slowly orbited a barren dirt knoll. Darkness now hid much of its features. To our surprise, we did not descend. Instead, the helicopters returned to our original PZ and we rejoined the company. Apparently, the chain of command determined that placing the unit on the hill at night would be too risky. We spent a restless night anticipating the next day's action. On Thursday morning, February 18, 1971, the helicopters picked us up again and headed back to the barren hilltop.

Alpha Company from the 4th Battalion, 3rd Infantry of the 23rd Infantry Division (Americal) clung tenaciously to that summit. Troops from the 27th NVA Regiment were attempting to control the high ground north and west of the Rockpile.[156] The ridgeline from which that hill projected formed a critical element in that strategy. Alpha Company had taken the hill on February 14 and operated from it for the next four days.[157] Someone alleged the hill's name was

---

156. Combat After Action Report 5th Inf Div Lam Son 719 CAAR (www.societyofthefifthdivision.com/vietnam/caar.html) transcribed by Kieth Short. Accessed March 7, 2018.
157. See Nolan, *Into Laos*, pp. 164-175, for an account of Co. A, 4/3rd Inf.'s com-

Nui Baí Lao. If not, it surely had a number on a map based upon its elevation. Whatever its official designation, those who fought there remember it by a different appellation. The men of Alpha Company, 4/3rd Infantry and Charlie Company of 3/187th Infantry call that red dirt hill "Purple Heart Hill." Charlie Company 3/187 Infantry was sent to Purple Heart Hill as replacements for the battered grunts of Alpha Company 4/3rd Infantry from the Americal Division.[158] The Rakkasans had our own "rendezvous with destiny."[159] Many of its grunts were to earn the hill's namesake decoration in the coming week.

The night had been long and ominous. Those grunts on last night's aborted CA couldn't shake the mental image of that barren hilltop. Enough helicopters were available by morning to convey the entire company in one move. So Charlie Company policed up its NDP, filled its foxholes with dirt, and redeployed onto the PZ. I was assigned to the fifth bird into our PZ. The birds landed at the PZ in five waves of four helicopters each. I would be in the second wave. The LZ was much smaller and able to handle only a single bird at a time. The spacing of the waves would facilitate landing one bird every two minutes in this unremitting arrival of fresh grunts. The existing

---

bat operations between February 11 and February 18, 1971.

158. Both battalions were temporarily detached from their parent division and were under the operational control of the 1st Brigade of the 5th Infantry Division (Mechanized) for the first phase of Dewey Canyon II.

159. On August 17, 1942, Major General William C. Lee, first commanding general of the 101st Airborne Division, issued a message to the division saying, "The 101st...has no history, but it has a rendezvous with destiny." [Rapport and Northwood, *Rendezvous with Destiny*, p. 3. "Rendezvous with Destiny" became the division's motto.

presence of the Americal grunts negated the usual risks of a single bird landing.

Charlie Company flew high above green hills covered in jungle vegetation or elephant grass. A light gray mist concealed much of the lower elevations. The vista resembled the view from dozens of previous combat assaults as we raced towards Purple Heart Hill.

When the choppers started to descend, they made a wide curving turn. This allotted me a better glimpse of our destination than we had the previous evening. A high ridgeline was visible in the distance. Its slopes were steep and covered in thick rainforest. At one end, a barren red dirt hilltop rose above the neighboring peaks. On the left side of the hill, the ridge appeared to drop rapidly towards the valley floor. On its right side, the drop was less dramatic. There the incline gradually leveled out into a series of squat knolls and shallow saddles that gave the silhouette of the ridge an undulating appearance. The rising and falling crest continued several kilometers before growing into another high barren peak. As the distance to the ridgeline decreased, numerous bomb craters on some hillsides became perceptible. It also was apparent that the sides of the ridge were extremely steep. Several dead tree trunks rose up from the crest of the barren hill. These lifeless trunks were devoid of branches and foliage. As we got closer, I could see a perimeter of foxholes ringed the first barren hill. A few low bushes were scattered over the hill. On its right side, just inside the perimeter was a small level area large enough for a single Huey. Yellow smoke billowed up from this dirt pad to mark our destination. A handful of grunts squatted at the edge of this LZ, ready to propel their bodies onto our choppers as swiftly as Charlie Company vacated them.

Our helicopters began a rapid descent onto the designated LZ. My seat on the helicopter floor afforded me with a spectacular view of the leading helicopters in the operation. The first chopper briefly hovered about a foot or so above the ground. Its prop wash caused the yellow smoke and red dust to swirl around like a multicolored lasso. Grunts jumped out of the bird and sprinted through the vibrant cloud to the edge of the LZ. An NCO started assigning vacant foxholes to them. Grunts from the 4/3rd Infantry dashed into the dust and smoke, and climbed onto the scrambling chopper. The second chopper took small arms fire as it approached Purple Heart Hill. Red and green tracers tore out of the jungle around the hilltop as the bird unloaded its precious human cargo. American and Communist tracers were not the same color. *Green... NVA. Red... USA.* The red tracers came from a captured M-60. *Not a good sign.* Puffs of grey smoke were clearly visible on the hill's summit. *Incoming NVA mortar rounds!* We really were going into a hot LZ, a very hot LZ. My heart was beating faster as my chopper descended into the inferno. *Nowhere to go but I've got to get out of this bird and find me some cover...*

The choppers barely slowed down as they hovered briefly on the LZ. Standing on the skids, we jumped hurriedly before the helicopters sped away to safety. As we tumbled to the ground, Americal grunts climbed swiftly into our vacated spots. The grunts made the switch rapidly since pilots did not pause long on the battered hilltop. Helicopters occasionally took off without a full load. This did not create a major logistical problem for the exchange. The battered defenders needed fewer spots than had been required to transport our fresher company. However, by the time Charlie Company's CA was completed, rumor claimed two of our grunts had been killed and 12 more wounded.[160] The chatter was worse than reality. Miraculously no one actually was killed.[161] A dozen or so grunts had been wounded however.

---

160. Nolan, *Into Laos*, p. 175.
161. No names from the 3/187 Inf. are on the wall for 18 February 1971. Wade

THE FLAG WAS STILL THERE

* * *

By late afternoon, the replacement of Alpha Company 4/3rd Infantry by Charlie Company 3/187th Infantry had been completed. Predictably, the impression each unit made on the other was exaggerated and misread. Charlie Company perceived the Americal soldiers as beaten and demoralized, which they were not. The Americal company identified the Rakkasan grunts as inexperienced rookies coming into the field for the first time without a clue as to what combat was like. One Americal platoon leader described the 101st Airborne grunts as wearing clean fatigues and black jungle boots.[162] Few grunts in Charlie Company had black jungle boots. We all took great pride in the doeskin appearance that the leather segments of the boots acquired in the bush. It was a visible sign that the wearer was a veteran soldier and most grunts in Charlie Company were veterans. In reality, both companies had been in the bush almost three weeks.

Three factors contributed to the misconceptions. First, Alpha Company (Americal) had been constantly engaged in heavy combat with the NVA for over a week. Their jungle fatigues were soaked in sweat and coated with dirt as a result. Their faces had acquired that 1,000-yard stare that comes from too little sleep and too much stress. The daily contact with a determined foe left little time for their normal

---

Rollins (Co E) died on 19 February and Edward Francis Downey Jr (Co C) on 21 February Downey was killed outright by small arms fire. (Virtual Wall full profile) He was killed in the mortar attack (see below). Were the dead from helicopter crews? No. The only names on the Wall from aviation units were entire crews from a Marine unit and from a 101st Avn helicopter that crashed on the border and was never recovered. It is likely that two seriously wounded casualties were rumored to be KIA.

162. Nolan, *Into Laos*, p. 173.

grooming. In contrast, Charlie Company (101st Airborne) had been operating in a comparatively quiet AO. Consequently our appearance naturally did not yet display the effects of prolonged fighting.

Second, the 101st Airborne Division still preserved a stricter uniform dress code than did many units in Vietnam at that time. Likewise, the Rakkasans still retained a strong battalion esprit de corps.[163] Although each soldier found some means to display individuality, pride in belonging to the battalion, to one's company, and to one of its platoons still exerted an influence on appearance. Rakkasan grunts kept their hair relatively short and except for mustaches, no facial hair was tolerated. Even in the bush, they routinely shaved. The logistics of resupply preserved some uniform consistency within each company. Furthermore, the brightly colored division patch on the left shoulder of most 101st grunts enlivened their uniform's appearance. Its conspicuous appearance eclipsed the stark olive and black patches worn by other units.

Third, members from Battalion Headquarters Company also participated in the CA. These soldiers may indeed have been wearing polished boots and clean uniforms. The unit contained more career soldiers and regulation uniforms mattered to them. Under their influence, even draftees in Headquarters Company conformed to lifer nonsense.

***

The company from the Americal Division had suffered extremely heavy casualties and appeared to us to be in "a disorganized state of

---

163. I cannot say what the dress code and esprit de corps of the Americal company might have been. I only know that in our battalion these still were extremely strong.

shock."[164] Rumors claimed the NVA had driven them along the ridgeline from abandoned FSB Scotch, discarding weapons, equipment, and their dead. Like most military gossip, the rumors were inaccurate. Alpha Company, 4th Battalion of the 3rd Infantry Regiment had not been driven from FSB Scotch. They had been fighting in the valley below and on the hills around Purple Heart Hill for eight days. On February 15, the NVA ambushed the company's first platoon. In the firefight that followed, the NVA established superior firepower and forced the platoon to make a tactical withdrawal, leaving the bodies of three dead comrades behind.[165] Over the next two days, the company made three separate attempts to recover the bodies. All three times extremely strong NVA resistance prevented Alpha Company from getting to the dead men.

Our first mission was to help recover the three American bodies lost in the ambush. As we retraced the route their RIF had taken a few days earlier, the sight of the cast-off equipment shocked us. An M-16 lay in the dirt. The fact that the NVA had not confiscated it suggested that the enemy was sufficiently supplied with their own weapons. They had no need to grab ours. Even more shocking was a discarded red mailbag resting against the base of a tree.

The three dead Americans already had been located and secured by their comrades in Alpha Company. As we moved to assist them, a brief but sharp firefight erupted. Two more men were slightly wounded. I was in the middle of the column when the shooting commenced. The distinctive sound of the AK-47 propelled me to the ground. Lying in the dirt, I wondered what was happening up front. This kind

---

164. A firsthand account of my tour in Vietnam written by me at Camp Evans and dated "5 Sep 71."
165. Benjamin Burgos-Torres, David Wayne Comber, and Julian Ernes Marquez (Panel 05W Vietnam Wall).

of uncertainty was one of the reasons that I preferred walking point to any place in the column. At least there I generally knew what was taking place. Word soon was passed up the line that all's clear and the recovered bodies were being sent up the column.

We held our position and the three corpses were passed up the hill head first. Each body was wrapped in an improvised poncho litter. The faces were covered but a pair of boots was visible at one end. Two men handed the front corners of the folded poncho to the next two in the line. After passing the front corners away, the men releasing them reached back and took the back corners from the men below them. Thus the three corpses were passed like water in an old fashion bucket brigade. As I extended to grab one end of the first poncho, the stench overwhelmed me. They had lay exposed to the elements for several days. The cold nights had slowed decomposition but the pungent odor of decaying flesh was still evident. I wanted to vomit.[166] Nevertheless, I braced myself and passed all three of the deceased warriors up the hill.

The platoon laid the bodies side by side at the edge of the LZ. Each was wrapped in a green poncho. With the heads covered, the jungle boots of each deceased man stayed exposed. On his way to his foxhole, Russ Kagy stopped and stared at their boots. He momentarily glanced back to his own feet. The boots on the three corpses were identical to the boots he wore. "*That could be me,*" he thought.[167] I also stopped briefly. For a fleeting moment I wondered about their families. My sanity prohibited such thoughts for too long. I clutched my

---

166. Nolan reported the cold night "had prevented them from decomposing" (*Into Laos*, p. 173). Russ Kagy, Alan Davies, and myself discussed this on February 9, 2018, at the 101st Airborne Division Snowbird Reunion in Tampa, Florida. All three emphatically remember a heavy stench from the bodies.

167. Russ Kagy, in a telephone conversation with the author, Monday, July 11, 2016.

M-16 by its slip ring and walked briskly to find Egg or Lieutenant Harris for my orders. Within an hour of recovering their bodies, the dead grunts were loaded on a UH-1H Huey and flown to the rear for shipment back to their families in the United States for burial.

There was no water on the barren hill. So the CO ordered first platoon to move down into the valley and locate a blue line. The hour already was late. Before much longer the jungle below us would block the last sun rays and transform the countryside into obscure darkness. The platoon cautiously descended single file down the steep slope. Unable to reach the valley floor before dark, the platoon set up a hasty NDP and began to dig in. However, army intelligence had learned of a planned NVA attack against one of the hills in the area. First platoon abruptly was ordered to hump back up Purple Heart Hill. Heavy rucksacks were packed and pulled up onto weary backs. It had been a long eventful day. Nevertheless, it would not be over until we reached the safety of the summit.

During the brief time first platoon had been stationary, the grunts heard loud primate sounds coming out of the valley below us. Members of first platoon still debate whether the noise came from monkeys or the mythological rock apes. Legend says the rock ape is a humanoid that inhabits isolated jungle regions in Southeast Asia. Reports say these muscular biped creatures are about six feet tall. Brown or black hair covers their body, except for the knees, the soles of their feet, their hands and face. These parts remain bare skin. Both sides in the war are said to have encountered these beasts. Most sources identify these creatures as myths, the Vietnamese equivalent to Big Foot. Many Vietnam veterans however claim that they are real. Some type of ape or monkey made the noises we heard that evening, but what kind likely will remain a mystery.

Step by step the grunts of first platoon slowly climbed up the steep hillside. Night already enveloped the American position on top of the hill when first platoon staggered into the perimeter. We were totally exhausted. Even so, the rest of the company shifted positions so first platoon could cover a section of the perimeter. We remained on 50% alert all night. Consequently the fatigued grunts only slept a couple of hours. Before dawn everyone was awake and in their fighting positions. The sun slowly arose over the infamous Rockpile.[168] All remained quiet. As the gray mist vanished and the landmarks around the hill clearly took shape, tension began to ease. Thoughts turned from combat to breakfast.

Hungry grunts pulled improvised stoves out from their rucksacks. Heat tablets were placed inside them. After rummaging through his various cans, the selected meal was opened with a P-38 by a twisting motion. The lid was left connected to the can, but was bent so as to serve as a handle. A match set the heat tablet ablaze. He then placed the open can of food on the stove. When the meal started to boil, the grunt lifted off the stove by the lid and stirred it with the plastic spoon from the brown condiment pack. After stirring, he brought a sip to his lips. If the taste was too cool, he returned it to cook a little longer. Otherwise, he quickly devoured the can's contents. A small roll of C4 plastic explosive replaced the heat tab. Its blue flame rapidly brought a canteen cup of water to a boil. Instant coffee was poured into the cup and stirred into the reviving beverage. Sugar and creamer packets were torn open and poured into the coffee according to individual taste.

The day was spent digging in on Purple Heart Hill. Previous bombardments had destroyed much of the vegetation on the hill. A perimeter of shallow foxholes already surrounded the crest of the knoll when we arrived. They enjoyed a wide, clear field of fire, but they needed strengthening. We dug them deeper and added overhead

---

168. Quang Tri Province, South Vietnam. UTM grid reference is XD912563.

cover. When completed, each position held four men. Larry Klag, Joe Sej, Recon, and I occupied a hole together. Our first efforts would have been comical were it not for the urgent necessity for shelter. Utilizing tree limbs lying nearby, we fabricated a bunker roof. However as we added dirt, it collapsed. The nearest logs were rotten. Thereafter, taking our weapons, we descended down the slope to secure some larger logs lying near the tree line. After dragging several logs up the hill, we placed wood crates filled with dirt on both front corners of the hole. A single eight inch diameter log spanned the opening between the two crates. It would support the new roof structure and provide a firing port. A smaller log was wedged in front of the larger log to stabilize it against lurching forward. These two logs were secured with sandbags. Two layers of logs reinforced the roof. The first layer was perpendicular to the support log. Its four inch diameter logs were cut to length with a machete. One end of each log rested on the support log. The other end was burrowed into the slope behind the hole. The second layer was perpendicular to the first layer. It was composed of short logs and cardboard tubes. Mortar rounds were shipped in cardboard tubes. A large number of such tubes lay discarded around our mortar positions. A sufficient number of these were confiscated and filled with dirt. Then we laid them out on with the logs. A layer of dirt was piled on top of these tubes and logs. Finally a layer of sandbags covered the entire bunker; When finished, our roof cover was over three feet thick. An opening on the front of the bunker allowed us a strong firing position. We set out several Claymore mines in front of the bunker.

An incident occurred that day that I should have recognized as providential. I was assigned to a detail from first platoon digging a slit trench adjacent to the hill's LZ. After a brief shift in the hole, a detail from another platoon replaced us. This rotation of diggers enabled those grunts on the detail ample time to prepare their own position.

When my replacement arrived, I handed him the shovel and climbed out of the trench. I picked up my weapon and headed briskly to my bunker.

As I approached my bunker I heard a muffled explosion from the direction I had just left. Initially I fell to the ground and lay there. When subsequent explosions did not follow, I recognized that we were not under attack. I jumped up and ran to see what had happened. Lying on the dirt was my replacement. He was covered in dirt and was bleeding in several places. A medic was working on him. His shovel had hit an unexploded plastic mine used by the Marine Corps earlier in the war. Fortunately, the ground had absorbed much of the blast and his injuries were not serious. The wounded grunt returned to his unit without being medevac'd. If he had been 60 seconds later in reporting, I would have been the one wounded that day. As for the hole, battalion flew in a mine detector and its operator. He scanned the area before work resumed. All the same, before sunset it was completed.

North Vietnamese soldiers were clearly visible on the hill where we discovered the mailbag the previous day. The CO requested air support. An old Douglas A-1H Skyraider prop torpedo-dive bomber of the South Vietnamese Air Force responded. The single engine Skyraider had been designed during the closing months of World War II. Despite being a prop-driven antique in the jet age, the Skyraiders competently fulfilled the infantry's need for close air support. Each plane carried up to 8,000 pounds of ordinance. This exceeded the bomb load of the iconic Boeing B-17 Flying Fortress.[169] United States Navy Skyraiders flew their last combat mission in support of the Marines during the 1968 siege of Khe Sanh. However, the United States and South Vietnamese Air Forces both still flew the A-1E and A-1H Skyraiders in 1971.

---

169. Don Hollway, "More Than Just A Prop," *Vietnam* 30 (August 2017): 27.

The rugged old plane's bomb run was not the screeching vertical nosedive shown in old movies, but it hit the target. Diving at approximately a 45-degree angle, it dropped heavy ordinance on the enemy occupied position. As the pilot pulled out of his dive, a colossal bomb separated from the plane's belly. Large fins popped out as the bomb plummeted earthward. A massive explosion left the hillside denuded and covered in debris. A huge crater marked where the bomb struck. Nevertheless, we could still see NVA soldiers scurrying about on the peak afterwards. The ineffective bomb run likely was due to an inexperienced pilot rather than to the equipment.

During the bomb run, the iconic old warplane captured the curiosity of the American grunts. The dark tan and green camouflage pattern copied that seen on United States Air Force airplanes. The national insignia was the same but enigmatically different. The blue roundel with a white star appeared identical, but the flanking bars were incongruous. Instead of the familiar white and red bars, they were yellow and red. In addition, a dark yellow and black diamond checkered pattern band encircled the fuselage just in front of the tail. But the plane itself was the real anachronism. The silhouette of its 50 feet straight wingspan and the vintage noise of its rotary engine were reminiscent of documentaries about the wars of earlier generations. They did not correspond to the persona of modern air support we customarily received.

The nostalgia abruptly ended when an 82mm mortar round slammed into the hillside below our position. Everyone lunged for the nearest cover. Some grunts dropped into nearby foxholes. Others just dropped into a prone position and hugged mother earth. Larry Klag had stepped outside the perimeter to answer nature's call. He was squatting over a small toilet hole he had dug when the round exploded. The explosion sent him scurrying back up the hill for cover. Some of those previously watching the Skyraiders now watched

Larry in astonished amusement. The snow-white skin of his thighs and his trademark sunglasses punctuated the incongruous humor as Larry attempted to run, carry his rifle and entrenching tool, and pull up his pants simultaneously.[170] The shelling that day did not last long. It served however to motivate additional improvements to our extemporized field bunkers. The defensive strength of our position on Purple Heart Hill was growing each day.

Friday afternoon, Battalion ordered the company to assault a nearby hill the next day. Saturday, February 20, 1971, would be Charlie Company's third day on Purple Heart Hill. The objective was located on the ridgeline that ran between Purple Heart Hill and FSB Scotch. It was only about 700 meters (less than one half mile) from Purple Heart Hill. A well-used trail on the crest of the ridgeline connected the two hills. The trail descended from Purple Heart Hill into a shallow saddle and then rose up to cross a small knoll. This knoll was where we discovered abandoned American gear on the first day. The Skyraider had bombed its eastern slope on the second day. The trail dipped off the knoll into another saddle before climbing up to the hilltop designated as the objective for the assault. During the week before we arrived on Purple Heart Hill, NVA repeatedly blocked every attempt by the American's Alpha Company, 4th Battalion of the 3rd Infantry, to take the targeted hill. It was during one of these attacks that the 4/3rd Infantry grunts were forced to abandon the three bodies we recovered the first day Charlie Company was on Purple Heart Hill.

Charlie Company's mission for 20 February was to drive the NVA off that hill. Needless to say, the grunts were less than enthusiastic for the assignment. In fact their attitude was nearly mutinous

---

170. This incident was shared with me by Alan Davies in a telephone conversation on December 28, 2017.

Friday afternoon as word of the mission spread. The two Kit Carson Scouts[171] assigned to Charlie Company absolutely refused to go on the mission. Nervous American grunts soon chimed in. "I'm not going up that d***ed hill," they muttered. The chain of command was breaking down and the company's combat effectiveness was declining. Individual soldiers showed selfish autonomous behavior rather than attention to preserving the unit. Preservation of the unit was not some mystical military sprite de corps. Rather it was a cohesion formed by a bond that put survival of other members in the unit ahead of one's own life. Living together every day for weeks created this bond. It came from sharing everything, good or bad. This bond was stronger than ethnic or regional identity. It formed a brotherhood that only those who have experienced it can understand. Now instead of thinking about how one's actions might affect one's comrades, people proposed an independent course of action without concern for other people's lives. The situation threatened to deteriorate into disaster.

Midafternoon someone popped smoke on the LZ and a shiny UH1 Huey with a Screaming Eagle patch painted on its nose touched down. Colonel Sutton darted out. To our disbelief, the Huey sounded as if it might remain on the pad. Leaving a Huey sitting still on the pad invited incoming enemy fire, lots of incoming! And our bunker was adjacent to the LZ where it was parking. Larry Klag swung his M-16 up into the firing position and aimed it at the chopper pilot. Holding the pistol grip with his right hand, Larry pointed up with his left hand. The pilot gave him the thumbs up sign and straightway lifted off. Sutton remained on the ground while his chopper circled irregularly at a high altitude.

---

171. Kit Carson Scouts were former Vietnamese Communists who had defected under the Chiêu Hôi program and agreed to serve with American combat units.

Before his chopper returned and he departed, Sutton spoke to the grunts designated to make tomorrow's attack. In his speech he boasted, "Boys we don't own part of a hill, we own all the hills around here!" Most of the men in Charlie Company thought his words were pure nonsense. They might have struck a chord a few years earlier when America still was fighting to win the war. But in early 1971, everyone knew the United States was leaving Vietnam without winning. The grunts did what was necessary in combat, but their motivation was to save the lives of their buddies, not bravado. The politicians in congress had already determined the fate of Vietnam. The grunts on the ground just wanted to get home alive. No one wanted to die for a losing cause.

Nevertheless Sutton's visit seemed to have a quietening effect. Talk of "not going up that d***ed hill" ceased. The insubordinate refusal to attack the hill likely was just normal G. I. whining. Although if it had persisted, some disillusioned grunts actually might have refused to fight. The repercussions could have been catastrophic. Instead, that evening a grim resolve permeated the atmosphere on Purple Heart Hill. "Not going up that d***ed" hill" no longer was an option. It never had been. Charlie Company was going up that "d***ed hill" whether or not its grunts favored going or not going. On a more positive note, Sutton promised two aircraft carriers, the U.S.S. Kitty Hawk and the U.S.S. Ranger, were offshore, ready to support our attack.

The battle plan called for first platoon to advance northward along the west side of the ridge. We were to move about 50 meters down from the crest and advance parallel to the summit. Second platoon was to travel parallel to first platoon on the east side of the ridge. Third platoon was to occupy the saddle between first and second platoons and adjacent to their objective. Its assignment was to lay down suppressing fire as first and second platoons wheeled to attack up the slopes. The two assaulting platoons were supposed to converge on the crest of the hill and occupy it.

THE FLAG WAS STILL THERE

During our preparations that morning, battalion headquarters advised Charlie Company that tube artillery could provide neither fire support nor prep fire because of a shortage of shells for the guns. Enemy sappers allegedly had hit the ammo dump at Quang Tri the previous night, cutting off the supply to the batteries firing in support of the American infantry. Until the destroyed rounds were replaced, all batteries were reserving their existing supply for emergencies. The news of this deficiency did not bode well for Charlie Company's assault. Close artillery support consistently gave American infantry an advantage in ground combat.

Sappers were elite NVA and VC assault and commando units. They were trained to infiltrate American bases and destroy high-priority targets such as ammunition and fuel depots. These select warriors easily penetrated the layers of barbed wire that surrounded Allied bases. They also spearheaded ground assaults by breaching the perimeter and attacking American defenses inside the target. Whereas a regular NVA infantryman received one month of training, sappers trained for six months. Their specialized training incorporated reading maps, breaching various obstacles, removing booby traps, handling explosives and demolitions, and digging trenches and tunnels. Their favored weapon was the satchel charge, an explosive device more powerful than a hand grenade.

To take the place of the missing artillery, the CO called in an airstrike before we moved out on the attack. The system used for tactical air during Dewey Canyon II insured no target was more than 15 minutes away from an airstrike.[172] A few minutes later, two Navy McDonnell Douglas F-4 Phantoms off one of the carriers in the South China Sea zoomed towards Purple Heart Hill. The Phantom was

---

172. 101st Airborne Division (Airmobile) Final Report, Airmobile Operations in Support of Operation Lamson719 8 February—6 April 1971, Volume II, p. IV-76.

the most formidable American fighter-bomber of the war. The mere sight of these aircraft roaring overhead produced a psychological lift in grunts on the ground. Their powerful payload could be a game changer in any dire situation.

Lieutenant Harris had reassigned me from point man to a platoon RTO for this mission. I felt honored to receive this assignment. Selection as an RTO showed superiors' confidence and earned peers' respect. An RTO was chosen on the basis of his combat experience, his coolness under fire, and intelligence. A good RTO would on occasion make command decisions for the officer to whom he was assigned. Constantly he acted to shield his officer from the ire of superior officers. Simultaneously, the RTO worked to enhance the confidence subordinates had in the officer he served. His responsibilities integrated the unit's administrative and logistical needs into his communication role. Reading maps was mandatory since he managed the unit's location, including all its dispersed elements. His biggest responsibility was maintaining communication with his unit's chain of command. Without this radio connection, the unit could not call for fire support, medevacs, or resupply. Consequently, RTOs were high-value targets for their enemy and the radio's antenna easily identified them on a battlefield. Although no increase in pay or promotion in rank was attached to my new role, the job did provide a certain status within the company.

I listened to the airstrike on my radio, a PRC-25. The Prick 25, as it was called, was a transistorized FM receiver-transmitter that weighed approximately 25 pounds. It had 920 channels and was the primary infantry platoon radio at the time. Two large knobs on the top of the radio controlled its frequency, which changed often in order to prevent the enemy from monitoring transmissions. A bulky battery provided power to operate the radio. These had to be replaced frequently, so an RTO always tried to have an extra battery in his rucksack.

THE FLAG WAS STILL THERE

The Phantoms made a preliminary pass in order to identify the target and nearby friendlies. The two Phantoms were so low as they whizzed over our bunkers that each aircraft's two crewmen were clearly visible through the canopy. I held the radio's black handset to my ears and listened carefully. The radio static this morning was louder than usual but I instantly recognized the CO's deliberate cadence.

"Roger. This is Charlie six, over."

"Affirmative, Charlie Six, this is"... the combination of radio static and the roar of jet engines momentarily drowned out the pilot's vibrating voice..."five. I have your paws in sight. Will make first run from the east, over."

By now the two aircraft had completed a wide loop and were approaching the target with the rising sun to their back. On their first run the two aircraft dropped high explosive bombs. Loud explosions rumbled across the countryside. One of the pilots chuckled over the radio.

"I have November Victor Alpha standing up and firing Alpha Kilos at me. Big mistake Charlie, over."

He spoke with an almost arrogant tone. AK-47s verses F-4 Phantoms was no contest. The two fighters banked and came back around, strafing the enemy position with rockets. The roar of their General Electric engines reverberated in my ears as both jets aimed their nose heavenward and climbed high, almost out of sight. Suddenly they rolled over, and descended in a screaming dive. As they pulled out of the dive, four elongated football-shaped canisters tumbled to earth, leaving the hilltop covered in burning napalm. A shout of approval thundered from the grunts on Purple Heart Hill. The two planes then circled our position and headed back to their carrier in the South China Sea. Black smoke still ascended from our target. We were left with our company's 81mm mortars as our primary fire support.

Nonetheless, the order to move out was given. The two Kit Carson

Scouts remained adamant. They sat on the ground near the perimeter and refused to venture off Purple Heart Hill. "No go. Beaucoup NVA, too beaucoup NVA. No go. NVA kill all Americans. No go!" they muttered repeatedly. Their anxiety was obvious. Yet Lieutenant Harris persisted in his effort to persuade them to go rather than assign point to a member of first platoon.

The entire company was clustered together in a tight column on Purple Heart Hill. Dreadfully conscious of the danger created by standing around in the open, everyone nervously waited for first platoon to move out. Finally an exasperated Sergeant Chester Lyons strode up from second platoon to check on what was causing the delay. He listened impatiently to Harris and the two Kit Carson Scouts. Finally he murmured, "To h*** with it!" He clutched his M-16 pistol grip and said distinctly, "I'll do it. Follow me." Lyons stepped over a log and headed down off the ridgeline. First platoon instinctively followed.

The platoon walked out of the perimeter in single file and headed down towards the first saddle on the main trail. Lyons walked point with Cecil Beach at slack. Larry Klag was the third man in the column. Harris came next with me in tow. Big Steve with his indispensable M-60 machine gun was behind me. Then his assistant gunner Uncle Al followed close behind him. The remainder of first platoon was strung out behind them, spaced about a yard apart. Kevin Owens walked drag.[173] Second platoon followed our drag man. Third platoon brought up the rear.

On this occasion, the grunts carried only the essentials for the assault: weapons, ammunition, and water. A few men carried their extra M-16A1 magazines in regulation cartridge pouches attached to a web belt. However, most carried extra magazines inside cloth

---

173. Kevin Owens, in a telephone conversation on January 10, 2018.

bandoliers hanging loose across the chest. For this attack, a majority carried two bandoliers. Each bandolier held seven 20-round M-16 magazines. With this system, retrieval of fresh magazines was slightly more difficult than from the pouch—but nearly twice as much ammunition could be carried.[174] Most grunts also carried one 100-round belt of M60 ammunition over their left shoulder. Grenades hung from belts, cartridge pouches, or stuffed in fatigue pockets. One or two quart canteens were affixed to web belts.

When first platoon reached the bottom of the saddle, it turned left and moved stealthily down the side of the ridge into the jungle below. Second platoon turned right. Third platoon continued moving along the crest of the ridgeline.

Along the west slope of the ridge triple canopy blocked sunlight and the floor was relatively free of thick underbrush. I clung to Lieutenant Harris with my radio, never more than an arm's length behind him. Its handset hung off the harness on my left chest, the radio volume cut down low so that transmissions were barely audible. My M-16 pointed skyward as I clenched the pistol grip in my right hand. A round was in the chamber and my right thumb cradled its selector lever. My left arm hung free, ready to grab the radio handset or the M-16's handguard.

We were barely out of sight of the saddle when Sergeant Lyon reported a small trail matching our planned itinerary. The column veered to the right and started moving along the trail parallel to the crest of the ridge. At this point, the planned assault on the enemy-held hilltop began unraveling almost immediately. Second

---

174. Each regulation small arms ammunition case held four 20-round M16A1 magazines. Only two cases were attached to standard issue web gear. This allowed only eight extra magazines (or 160 rounds), whereas the bandolier system permitted 14 extra magazines (or 280 rounds).

platoon had veered off the saddle opposite first platoon but stalled in extremely rugged terrain. The constant bombing and artillery strikes had produced a maze of craters and fallen trees along their assigned route. Third platoon also fell behind as it worked its way through these same type obstacles. On the other hand, first platoon advanced unimpeded along a high-speed trail. Enemy trails posed numerous dangers. So we warily moved along on it. The point team walked with caution. Each step was deliberate and noiseless. Their eyes scanned every rock or leaf for clues of enemy presence. Notwithstanding, first platoon made excellent progress in its advance toward the mission's objective.

The underbrush grew thicker but the trail allowed us to keep an even pace. Just before reaching the point where we were to turn and sweep up the ridge, Lyons and Cecil discovered our path intersected another well-used trail that led towards the crest of the ridge. It was unclear if it went to the saddle — from which Third platoon was supposed to lay suppressing fire — or to the hill that was our objective. Lyons raised his left hand and the platoon froze in its tracks. Commo wire was laid out along the new trail's shoulder. There were no civilians in the area and the United States Army no longer used field telephones. Therefore, these telephone lines were part of an important NVA communication network. Lyons and Cecil promptly headed up the hillside on the new trail. The LT instructed me to stand fast. Then he and Larry started after Lyons. However, Larry stooped down and cut the wire with his knife before going. This essential action delayed him and Lieutenant Harris briefly. Meanwhile Lyons' precipitous progress separated Cecil and him further from the platoon. Next LT and Larry likewise moved out of sight but detached from the point team by Larry's cautionary pause. At the tail end of the column, Kevin felt a tug beneath his boot. Looking down, he spotted another wire stretching down the hill. It had been covered with dirt as the platoon crossed ahead of him. He pulled his knife out and cut the wire with a sawing action.[175]

As I stood waiting for the LT's return, I spotted a man's head about ten yards down the main trail and just off it in the underbrush on the far side of the trail. He was largely concealed by the bushes, but a slight movement of his head caught my attention. His khaki soft cap differed from the bright green vegetation.

I turned to Big Steve and softly inquired, "Who went down the trail?"

He responded emphatically, "No one. Everyone had gone uphill, not downhill."

"Are you sure?" I whispered.

My heart was pounding as I realized the man in the bushes was…*NVA! Calm down. Don't panic. Remember your training and do what you're here for…*

Big Steve's soft voice answered, "Yes. Lines, See cile, Lar ee, and the el tee went up that trail." Then he pointed to the direction Lyons, Cecil, Larry, and Harris had gone by nodding his head.

"Look ten yards down that trail in the bushes on the far side. Do you see anyone?"

"No…wait. Yes, he's wearing a khaki cap."

"Affirmative."

Quietly word was passed back that we had NVA to our left front. The platoon leader was elsewhere and there was no time to locate him. The platoon NCOs were solid and they seized the initiative. Squad leaders signaled their men. Heads nodded. Every man knew his job. The tension of anticipation rose. All eyes seemed to be looking

---

175. Kevin Owens, in a telephone conversation with the author. January 10, 2018. At the 101st Snowbird reunion in 2018 Russ Kagy remember Kevin sawing a wire with his knife.

at me, but only a few grunts up front actually could see me. My thumb pushed the selector lever down. It locked in the semi-automatic position.

My rifle range training kicked in. *BRAS* (pronounced brass). *Breath. Relax. Aim. Squeeze.* I wasn't conscience of it, but it controlled my action. My heart was beating so hard now it seemed to drown out every other sound. I took in a deep breath and momentarily released my body's tautness. I took careful aim at the NVA soldier, lining up the front sight post with his head through the rear sight. I squeezed the trigger. My first shot was slightly high. The round hit the back of his head and then went into his prone body. He emitted a painful yelp and began a mournful crying. Listening to the wounded enemy soldier's moaning was depressing. I put another round into him. This shot killed him and he ceased his sorrowful groaning.[176] I didn't see any other NVA soldiers. So I started firing into the undergrowth, putting my rounds about six inches above the ground surface.

Meanwhile as the old idiom says, "All hell broke loose." For a moment I heard the comforting cadence of Big Steve's 60. Tat tat tat tat tat. It was chewing up the leaves beyond the NVA I had just killed. Uncle Al was feeding it ammunition when he heard a sickening click. The gun had jammed. He and Big Steve worked frantically to

---

176. My memory erased the wounded man's groaning. I could remember shooting him more than once, but did not remember why I did. Over 45 years later in a telephone conversation, Russ Kagy mentioned his dreadful moaning. Russ's account revived the suppressed memory. Shooting the NVA soldier a second time now made sense. I shot him to eliminate the dreadful sounds he was making. Not only was it psychologically depressing; it indicated he was still alive. So I didn't know if he also was capable of inflicting harm on my platoon. Silencing him was the only means to insure he could not hurt any of us.

clear the weapon and get back into the fight.[177] Explosions from the M-203's 40mm grenades also were tearing up the jungle in front of us. But there also were more ominous sounds as well. The distinctive kak-kak kak-kak of AK-47s could be heard amidst the symphony of firing. These were punctuated by the base explosions of the big round Chinese equivalent to the Claymore mine. An RPG zoomed overhead and exploded harmlessly against the hillside with a dull boom.

Caught in the frenzy of the battle for a brief second we stood erect and poured suppressing fire at the unseen enemy. Dropping to a prone firing position, American grunts pulled new magazines from bandoliers suspended around their chests. They pushed the magazine button to release empty magazines from their weapon. Then they snapped the full magazine taken out of the bandolier into the magazine housing of their M-16s. As the metal magazines clicked, the heel of their dirty hands rapped the bottom of the magazine to insure it was in position. Quickly they pulled back the weapon's charging handle and released it. The bolt settled into position, chambering a fresh round. Index fingers squeezed triggers and M-16s spit their lethal missiles into the vegetation hiding an unknown number of NVA.

The men at the front of the column continued blasting away. Other members of our squad rushed forward and joined the combat. They swung out to our left and came on line beside us. They added more firepower to the melee. In the lingo of naval warfare we had "crossed the T." The full weight of our fire was sweeping down the NVA flank and beyond. Because they were perpendicular to our battle line, the NVA were unable to fire all of their weapons directly at us. Therefore, their firepower had minimum effect. Still enemy fire to our front increased momentarily. Then promptly it ended completely. By then

---

177. I didn't realize the M-60 had jammed until Alan Davies ("Big Al") revealed it to me in a telephone conversation in 2017.

everyone was on the ground or behind tree trunks. The grunts on the left flank saw an NVA soldier jump up and race down the trail towards the valley floor. He fell down, rolled over, regained his upright stature, and disappeared into the jungle foliage. His companions likewise bolted without being seen by us. Before the rest of the platoon could come on line, the firefight was over.

We had stopped to cut telephone lines and the stop saved us from disaster. The NVA apparently had set up an L-shaped ambush on the trail we were walking. They had deployed small arms and claymore mines. Several 82mm and 60mm mortars supported their careful laid trap. If we had continued walking just a few more yards, the NVA would have caught us in a crossfire of 12.7mm machine gun, AK-47s, RPGS, mine blasts, and mortar rounds. Very few of us would have made it out alive. However, one man's movement had revealed the trap and the predator became the prey.

We had flanked the NVA position and our firepower was too much for them. When one leg of the ambush gave way, the other leg became untenable. The NVA disengaged and withdrew into the jungle. Enemy casualties were unknown. The only body remaining on the field was the soldier I had shot. The NVA could not get to him without exposing themselves to our deadly fire. There was blood and evidence that we had hit others, but there was no way to determine how many or how serious their wounds. Like us, the NVA removed their dead and wounded from the battlefield. It was extremely important in their culture that the dead receive proper burial. They widely believed if dead bodies were left on a battlefield their spirit would wander forever. Therefore the North Vietnamese always carried off their dead if possible.

As soon as the hostile fire fell silent, Larry and the LT returned. They moved slowly down the trail to the NVA body.

While Larry and the LT checked the dead NVA soldier, Colonel Sutton called in on the company radio net. I answered, "This is One

Six Romeo, over." Adrenaline had me pretty hyped. I spoke rapidly as I summarized what happened.

"Charlie One Six Romeo, put your six on, over."

"Wait one, we've been in contact. One Six is checking November Victor Alpha dead, over."

"Say again, you have contact with November Victor Alpha dead? Verify, over."

"Affirmative, over."

"Give me a sitrep, over."[178]

"This is One Six Romeo. Roger. Almost walked into an ambush! We found wires on a trail. While we were cutting the wires we spotted Dinks hiding in the bushes. We opened up and blew 'em away. Over."

I must have been pretty excited and was still talking rapidly. Sutton responded, "Calm down, son. Speak slower. Over."

"Roger, over."

"Do you have enemy body count? Over."

"Negative. Over."

"One Six Romeo, tell Charlie One Six Actual to call in that body count ASAP. Understand, son? Over."

"Roger. Over."

"Any friendly casualties? Over."

"Negative, none to my knowledge, over."

"Very good." Sutton paused briefly and then asked, "Charlie One Six Romeo, say again, how did Charlie One discover that Gook ambush, over?"

"We were stopped, cutting the Dinks' telephone lines when I saw his head move. He was camouflaged pretty good, but that movement of his head gave him away..."

---

178. A situation report.

Sutton interrupted, "One Six Romeo, did you say 'you were the first man to spot the Dinks'? Over."

"Roger, it was me, sir...over."

"Good job. You have earned a three day Romeo and Romeo at China Beach, over.[179]

"Thank you, sir, over."

"Now, son, as soon as Charlie Six returns, tell him I want him to send in a sitrep and a body count. Over."

The sharp disparaging tone had vanished. Sutton's voice now sounded almost like a coddling grandfather. The report of enemy dead without American casualties likely portended a successful operation in his mind. Whatever caused this change was okay with me. I didn't like the battalion commander reprimanding me. Besides, I was going to China Beach... provided I ever got off Purple Heart Hill alive. Every soldier was entitled to a seven-day R&R during a one-year tour of duty in Vietnam. Three-day R&Rs were generally rewards for specific reasons. China Beach was an in-country R&R center on the South China Sea northwest of Da Nang

"Yes, sir! Thank you, over."

Larry and the LT returned while I was talking with Sutton. They brought back some papers, the dead man's web belt, his khaki soft cap, and his AK-47. The Lieutenant now was standing in front of me, listening. I handed him the mike and he briefly reported the action up to that time.

Larry handed me the belt, hat, and AK for souvenirs. The cap reminded me of a World War II Japanese hat. The belt was pale green webbing with a silver buckle. A star was engraved in the buckle. Later that week I photographed Larry Klag wearing the two items. I kept the belt, but the cap was somewhat gory from his head wound. Larry wanted

---

179. Romeo and Romeo was phonetic alphabet for "R&R," rest and recreation.

it so I gave the cap to him. The AK-47 had a folding stock and two 30-round banana clip magazines taped together with black electrical tape. I removed the magazines and ejected the round from the chamber. I also took the top round out of the magazine. These two bullets were more than souvenirs. They would serve as *tangible reminders of what took place on that hillside, copper testimonies that I had survived this day*. One of the papers identified the dead soldier as 33 years of age. Later that day we sent these papers and the AK-47 back to Camp Evans.

The firefight did not end our attack. Our objective remained the hilltop on the ridge above us. The other two platoons still had not yet gotten into position to support us. First platoon had little choice but to attack alone. The firefight had revealed our presence to the enemy. Remaining stationary invited our destruction. Furthermore, Lyons and Cecil had not returned to first platoon's location. It only took a couple of minutes to organize the assault. Larry took point and the platoon moved cautiously down the trail that paralleled the crest of the ridgeline. Then we made a right face. Lieutenant Harris and I stepped back behind the line. Big Steve closed on Larry. Harris bellowed, "First platoon, move out!" The line advanced. The pace was methodical but not impetuous. As we ascended, we emerged into open ground. I blinked my eyes rabidly. They quickly adjusted to the bright sunshine after the dark shade of the jungle.

American bombs and artillery shells had pulverized the location in the past. Now the ground was rocky and devoid of all plant life. A few isolated shallow craters were still visible, but these were only a few inches deep. I was out in this clearing when we heard the deep hollow thump, thump, thump of mortar rounds being dropped into mortar tubes. We had little doubt but that we were their target. Instantly everyone dropped flat. I pressed my body as firmly to the ground as I could. Then I pulled the radio up over the back of my head and prayed. I prayed real hard… I mean real hard.

The mortar is an indirect fire weapon. Its rounds have a high parabolic trajectory. The seconds between hearing them launch and seeing them explode are moments of intense terror. Exposed in the open as we were, you are totally helpless. There was absolutely nothing you can do to protect yourself. There was no place to hide. The insubstantial bomb craters offered an illusion of cover from direct small arms fire, but not mortar rounds raining out of the sky. You cannot fight back. The weapon in your hand cannot hit the unseen enemy. You must just lay still and wait... wait to see if you live or if you die. Russ Kagy summed it up succinctly when he said, "I've never been more afraid in all my life."[180]

I had time to say a lot to God in the time the mortar rounds arched across the sky. I was 20 years of age and had my entire life in front of me. I begged God for my life, giving Him every reason why I should not die. The mortar rounds struck just below the crest of the hill where I lay. Blam! Blam! Blam! I glanced up and saw gray smoke starting to dissipate. The last dirt clods were raining down on the slope above us. A few hit Larry and the LT, but not me. *I was alive! I had no wounds...* I gave thanks.

I told God, "Thank You." Then I heard that terrifying resonating thump, thump, thump once again. At least three separate locations were firing mortar rounds at us. My praying now grew desperate. I pleaded frantically to live. *I do not want to die in Vietnam. I don't want to die!* My prayer was ended by a salvo of explosions as the mortar rounds crashed into the ridge above me. Blam! Blam! Blam! I lifted my head to see smoke ascend from a location midway between my position and the spot where the first salvo had hit. Dirt clods showered down on me this time. I felt every pebble and grain of sand strike my body. As the rain of dust ended, I realized I was still alive.

---

180. In a conversation with Alan Davies and the author, February 9, 2018.

Simultaneously I breathed a sigh of relief and uttered a quick prayer of thanksgiving. *I was still alive and unhurt! Oh God! Thank you!*

One more time however my elation abruptly ended. For the third time I heard that unmistakable hollow thump, thump, thump! Instantly I recognized the situation. The NVA were walking their mortar rounds down upon us. It was a very sound tactic. We had wrecked their ambush, but in doing so the platoon had wheeled and started uphill. Because the ambush party had broken off contact with us, the NVA probably could not be certain where we were. Our actions during the firefight, combined with an American tendency to seek the high ground, indicated to them that we were somewhere between the ambush site and the crest of the ridge. By walking their barrage down the hill, they were certain to hit us. *Oh God, I'm dead for sure...*

In that moment I felt a peace I had never known previously nor experienced since. As the third salvo of mortar rounds winged their way through the air to kill me, I prayed once again. This time my prayer was markedly different. I no longer asked for life because a lethal shell already was on its way to take my life. Nothing I could do would stop it. In that moment in my mind, my destiny was fixed. I accepted my fate completely. In a few seconds I would be dead or dying. *No, I already was dead. My body just had not yet ceased its biological functions.* I simply said, "Lord, I know I have done some things that displease You. I regret what I did, but I do know where I will be standing in one moment." *I will be standing before Christ my Savior to be judged by Him. However, I have trusted in the good news of Jesus Christ's death and resurrection. I believe that faith in Him gives eternal life. My name is recorded in the Lamb's Book of Life. I must give an account for my sinful behavior, but I will enter into heaven and spend eternity with God. Of that I am certain.*[181]

---

181. The Bible says For you are saved by grace through faith, and this is not from

In that instant, God spoke to me. I cannot say if I heard a literal voice or if a thought was impressed upon my mind. The particular mode is irrelevant. I do know unequivocally that God communicated directly with me. He made two assertions. First, He identified Himself as the God who hears and answers prayers. He said that my church was praying for me to live and that He was going to grant their petitions for my life. I would live. Second, He stated that He had unspecified work for me to do in the future for Him. Time had run out. But this time the shells did not come screeching down on top of me. In fact they missed me so completely that I have no idea where they hit. Others may seek a rational explanation for their miss, but I believe God simply grabbed those rounds midair and hurled them elsewhere. The men who fired those mortars were very good. They arguably were the best mortar crews in the world. They would prove just how good they were the next day. They knew approximately where we were and they employed a sound tactic for the situation. I find the idea that they missed their target illogical. Only divine intervention satisfies the circumstances.

For the rest of my tour, I kept the round from the chamber of

---

yourselves; it is God's gift (Eph. 2:8 [Unless otherwise noted, all Scripture quotations are taken from the Christian Standard Bible®, Copyright © 2017 by Holman Bible Publishers. Hereinafter cited as CSB. Used by permission. Christian Standard Bible® and CSB® are federally registered trademarks of Holman Bible Publishers.]). On a Sunday evening in September 1959 at Immanuel Baptist Church in Lakeview, Georgia, Reverend Frank Phillips preached from the passage And brought them out, and said, Sirs, what must I do to be saved? And they said, Believe on the Lord Jesus Christ, and thou shalt be saved, and thy house (Acts 16:30-31, KJV). Later that night, believing that Christ died for my sins and rose on the third day to give me life, I knelt and asked Jesus to save me. Rev. Phillips was a disabled World War I veteran. He had been blinded by a German gas attack in France.

the dead NVA soldier's AK-47 in my pants pocket. I joked that it was "the bullet with my name on it." In reality it remained a concrete reminder of how close to death I came that day. I never again thought that I would die in Vietnam. I would have a number of narrow escapes in the future, but I believed God had promised me that I would survive. I continued to carry the bullet for the next 25 years. I finally decided that carrying an old Chinese Communist bullet in my pocket was unsafe. I stopped the ritual in the mid-1990s. I still have the bullet. It is displayed on a desk in my home along with a few other mementos from Vietnam. I see it daily and remember. Periodically I pick it up and examine it closely. When I do, I relive those moments of terror…intuitively I bow my head and thank God that I am alive.

Although I did not recognize it at the time, the experience on that hillside had been the pivotal moment of my life. Eight years later, God brought back the vivid memory of that moment and called me into the gospel ministry. Without hesitation, I knew that he had spared my life on Saturday, February 20, 1971, for the sole purpose of serving Him. Accordingly I became an ordained minister and served as a pastor for over 30 years. Since then I have served as an adjunct professor for New Orleans Baptist Theological Seminary training others for Christ's service.

When the final mortar rounds passed, the platoon arose and charged swiftly towards its original objective. Several large trees and patches of thick undergrowth crowned the hilltop. By now the enemy had abandoned the hill, but a horrifying sight greeted us as we reached the top. We finally linked up with our point team. One man was dead and one was seriously wounded. Sergeant Chester Lyons had been killed by a large Chinese claymore. He wasn't in first platoon, but had volunteered to go with us on this assault. He and Cecil had reached the crest of the hill. Lyon's charred body hung in a tree where the blast had blown him. The blast had mangled his head severely. Someone removed their jacket and tenderly wrapped it around his head. Lyons, age 23, was from San Jose, California. He was midway through his tour.

Nearby sat Cecil. He apparently had been hit with pellets from the same blast that killed Lyons. He was staring numbly into space and bleeding in numerous spots. Every medic in Vietnam was called Doc. Cecil tried to call for our medic. "Doa uha ga, doa uha ga," he muttered. The blood gurgling noise in his throat made his breathing add sound to his call for "Doc." Instead of a crisp "Doc," his plea sounded like a mumbled "doctor." Uncle Al reached him first. He cradled Cecil's head in his arms, seeking to comfort him. One hole in his neck overshadowed his other wounds. The injury had been caused by a large piece of shrapnel or an AK-47 round. Fortunately, it had missed the artery, but he still was bleeding profusely. Al looked on helplessly as Cecil turned white and plunged into shock. Doc raced up and immediately started to work on saving him.[182]

We needed to secure our objective in order to medevac Cecil safely. A perimeter was set up. Everyone checked ammunition and weapons in case the NVA counterattacked. Canteens touched parched lips as sweaty warriors swigged a few drops of needed water. Lyons' body was retrieved. I called in a Dust-Off to medevac Cecil.

Less than 15 minutes later, a UH-1H Huey approached the hilltop. The familiar red cross on a white square was painted on its nose and door. Larry Klag "popped smoke" by pulling the pin on a colored smoke grenade. These cylindrical grenades came in red, yellow, green, and purple. Red was reserved for marking targets. Infantry units marked their location by setting off one of the other three colors. Larry popped yellow. The grenade emitted smoke for several minutes after igniting.

---

182. Discussion between Alan Davies, Russ Kagy, and myself at the 101st Airborne Division Snowbird Reunion in Tampa, February 8-10, 2018.

THE FLAG WAS STILL THERE

I squeezed the switch on my handset and said, "Dust-Off eight five, this is Charlie One Six. Smoke out, over."

"Charlie One Six, Dust-Off eight five. I have yellow submarine, over," the pilot replied, his voice oscillating from the engine vibration.

SOP called for the unit popping smoke not to identify the color. The NVA sometimes also popped smoke in an effort to lure the chopper into an ambush. Therefore the pilot identified the color in order to confirm he landed in a friendly LZ. Pilots often labeled the colors with popular songs of the era such as "Yellow Submarine" by the Beatles.

"Dust-Off eight five, Charlie One Six, roger that. We have yellow smoke. LZ too small for landing. Use jungle penetrator. I repeat, LZ too small for landing. Use jungle penetrator. Do you copy? Over."

"Charlie One Six, roger. I have your ground marshaller in sight, out."

Larry raised his rifle over his head and guided the pilot into position amidst swirling colored smoke and dust. The terrain and the trees precluded it landing.

A jungle penetrator dropped down to us. The jungle (or forest) penetrator was designed to evacuate casualties when a LZ was not available or when the vegetation was too dense. It is 34 inch long cylinder that weighs 21.5 pounds. Three spring-loaded folding legs function as seats.[183] The bright yellow penetrator struck the ground, discharging the static electricity built up during its fall. A couple of grunts assisted Doc and carried Cecil to the penetrator. Two of its legs were unfolded and Cecil sat straddle of them facing the penetrator's main drum. He quickly was secured with the penetrator's webbing.

---

183. Aeromedevac summary sheet atzb-rcg, Department of the Army U.S. Army Pathfinder School the Army National Guard Warrior Training Center Fort Benning, Georgia 1 October 2009.

Doc looked up at the Huey crew and signaled them to retrieve the penetrator. The Huey's hoist rapidly reeled in the penetrator. Cecil was pulled up into the helicopter's cabin. The Medevac's rotary blades whipped up additional dust as they struggled to lift the bird skyward. Then with the familiar whoop-whoop sound it vanished from our sight. The whole process had only taken a few minutes but the time the Medevac hovered overhead seemed like an eternity for both the aircraft crew and the grunts on the ground.

Medevac pilots routinely risked their lives to extract the wounded. During the Vietnam War, air ambulances medevac'd between 850,000 and 900,000 casualties, including civilian and enemy wounded.[184] Casualties were classified "urgent," "priority," or "routine." Urgent cases were those who would die or lose a limb without immediate medical evacuation. Priority meant the patient was critical, but stable and expected to remain stable for up to four hours. All other patients were considered routine. Regardless of classification, patients frequently were in surgery within an hour of being wounded.[185] The average time it took to move a wounded soldier from the battlefield into a hospital was only 2.8 hours.[186] The grunts did not necessarily know the statistics, but they were keenly aware that it was historically fast. The confidence that if wounded they would be medevac'd rapidly for medical treatment supported good morale in combat engaged soldiers.

By this time, second platoon also reached the objective. Their first instinct was to secure Sergeant Lyons' body. They carried his corpse back to the LZ on Purple Heart Hill. Along with the dead NVA's papers and AK-47, his remains were flown back to the rear later that

---

184. *Encyclopedia of the Vietnam War*, 2000 ed., s.v. "Medevac" by Arthur T. Frame, p. 258.

185. Dunstan, *Vietnam Choppers*, p. 140.

186. *Encyclopedia of the Vietnam War*, s.v. "Medicine, Military" by McCallum, p. 261.

evening. From there, Lyons was sent to the U.S. Army Mortuary at Tan Son Nhut for embalmment and shipment back to his family in the United States.[187]

The initial contact had occurred at 1245 hours when I shot the NVA soldier. The last enemy fire was a few incoming mortar rounds during Cecil's Medevac extraction. That was at 1550, over three hours later. The hill that eluded American forces for nearly a week was now in our possession. Second platoon did not remain on the hill long. After taking possession of Lyons' body, they returned to Purple Heart Hill. Along the way, they linked up with third platoon. First platoon remained on the hill another half-hour or longer. We set up a perimeter. Then we neutralized several booby traps and Chinese claymores. Finally, first platoon also returned to our foxholes on Purple Heart Hill. Dusk was beginning to darken the hillside when we got there.[188]

THE STAR SPANGLED BANNER

The NVA retaliated the following day. For approximately five hours

---

187. Scuttlebutt in second platoon incorrectly claimed Lyons walked point because everyone in first platoon refused to do so. The only people to refuse to walk point were the two Kit Carson Scouts. Lyons recognized the extreme danger waiting posed and seized the initiative to shatter the stalemate between Lt. Harris and the Scouts. The same rumor mill alleged first platoon also refused to carry Lyons' corpse back to Purple Heart Hill. Such an accusation is absurd. The decision for second platoon to carry his body was made by the company commander, likely in response to a request by second platoon's platoon leader.
188. Telephone conversation with Russ Kagy (Monday, February 17, 2020); Transcript of 1/5th Infantry (Mech.) CAAR [Combat After Action Report] (http://www.societyofthefifthdivision.com/vietnam/significantevents.htm)

or longer they poured 60mm and 82mm mortar rounds and RPG rockets down onto Purple Heart Hill. We hunkered down inside bunkers and waited for the shelling to stop in anticipation of a massive ground attack. Our bunker was approximately six feet broad by three feet wide and five feet deep. We had covered it with three feet of logs and dirt. An opening in the front provided a firing port. Another opening on one side provided an entrance and exit. I was sitting next to "Recon" (Hector Torres). Larry and Joe sat on the other side of Recon. Our shoulders pressed hard against each other in the cramped quarters. An endless reverberation of explosions wreaked havoc on the hilltop.

One round exploded about 10 or 15 yards in front of our bunker. I sat mesmerized by a rotating piece of metal sailing directly at Recon and me. For an instant it appeared to spin in slow motion as it flew towards our faces. A split second later a thud signaled the end of the shrapnel's flight. The shrapnel—a shell fragment approximately a foot long and one inch wide by a quarter of an inch thick—easily could have cut both of our necks. Instead it split the difference between our ears and stuck into the back wall of our bunker like an arrow. Recon was a short timer, with less than three weeks left on his tour and the incident unnerved him significantly. He would not have been human if it did not. I wasn't exactly "calm, cool, and collected" myself. Recon pulled himself slowly together. A few days later he left the field and returned to Camp Evans where he completed his time in Vietnam before returning home to Puerto Rico.

During the prolonged shelling, Medevac helicopters made numerous sorties under fire in order to pick up the seriously wounded. I witnessed numerous anonymous heroes that day. They carried their buddies to the birds, knowing that in doing so they risked death or dismemberment. One Medevac lifted off and made a hard right bank to avoid incoming. When the helicopter turned, one wounded soldier rolled out and fell nearly 40 feet to the ground. He landed in front of Carl "Snuffy" Booth and Russ Kagy's bunker.[189] Snuffy bolted out of his protected hole and sprinted to the victim, a sniper from Company E. Concurrently, another grunt likewise dashed to the casualty's assistance. The soldier who fell was still alive. His two rescuers picked him up and carried him back to the LZ. Snuffy held the injured man in his arms until the Medevac could circle back and retrieve him. Snuffy did not know the injured man personally. To Snuffy he was an American soldier and that was all that counted.

We could identify at least three NVA mortar tubes firing at us. Charlie Company's mortar returned fire. The mortar crew knocked out one NVA tube, but shortly thereafter NVA rounds put our mortar out of action too. Inside their improvised bunkers the company's grunts continued to hunker down and wait as the shelling continued. We felt helpless. Our instinct alternated between the urge to fight back and a desire to run. We lacked the means to strike back, so we could not fight. There was no place to run, so flight was out of the question. All we could do was stay put and wait for the barrage to end. Waiting under fire during a bombardment was terrifying. Every second and every explosion amplified the terror. *Will the next round hit me? Will the dinks attack when the mortars stop? Am I going to die on this God forsaken hill? NO! Yesterday God promised me I will live.* Nevertheless, I was not able to escape the terror.

Charlie Company lost one man killed—SP4 Edward F. Downey Jr.—and 17 wounded in the shelling that day. Downey, age 21, was

---

189. Russ Kagy recounted this incident in a telephone conversation with me on Monday, July 11, 2016. Like me, the bravery of our comrades in arms greatly impressed Russ. For him courage is personified in the actions of his close friend Carl Booth. At the time neither Carl nor Russ knew the identity of the man who fell from the Medevac.

from Cleveland, Ohio. He had five months left until his DEROS date.[190] Battalion headquarters also was located on the hill. We understood that they had one killed[191] and six wounded.

Three dead trees stood on the summit of the hill. Practically all of their branches were gone, giving them the appearance of antiquated telegraph poles. Sometime during the shelling, someone in battalion headquarters ran an American flag up on one of the trunks. In the chaos of exploding shells, he attached it to the shortest tree trunk. When the shelling ended, he transferred the flag up to the pinnacle of the tallest trunk. It caught the breeze and billowed out against the backdrop of a bright blue sky. I have never been more patriotic than when I first gazed upon that defiant red, white, and blue banner. I know how Francis Scoot Key must have felt as he first glimpsed the Star-Spangled Banner flying over Fort McHenry in Baltimore Harbor during the War of 1812. I cannot hear the National Anthem today without recalling that flag on Purple Heart Hill. Russ remembered having a similar emotion. He later wrote, "Cannot begin to describe how proud I felt seeing the Stars & Stripes flying over our heads. It is something I will never forget."[192]

The National Anthem's first verse raises the question as to whether

---

190. His DA2496 form simply states "Individual was killed while on a combat operation when a hostile force was encountered."

191. No one from HHC of 3/187 Inf is listed on the Wall. Roger Maki of HHB 1st BN 321st Field Art was killed in Quang Tri Provence by mortar fire on this date. Was he an FO attached to HHC? The 3/187 Inf was in the 3rd Brigade. This battery normally provided support for 2nd Brigade, 101st Airborne Division rather than the 3rd Brigade.

192. Facebook website We Were Soldiers Vietnam (https://www.facebook.com/search/top/?q=photos%20in%20we%20were%20soldiers%20vietnam&bs=photos-in(303988719800321)&filters_rp_author=me.) Accessed September 11, 2015.

the republic will survive.[193] It survived the catastrophic invasion of 1814 and it has to date survived the Vietnam War. It will continue to survive only as long as its citizens are willing to fight and die for its existence as a free nation. Keith Nolan described the soldiers who fought in Dewey Canyon II as "perhaps the least motivated in U.S. history."[194] For a time on that barren hilltop, this was not the case. When the barrage ceased, guys stood out on the perimeter hurling taunts at an unseen foe below us. A substantial NVA assault could have taken the hill that evening, but the enemy would have suffered heavy casualties in doing so. Charlie Company, for once, was itching for a fight, an opportunity to pay back the dinks for the horror we endured during the five hours inside those bunkers. But no attack was launched. Instead, the NVA hit another Rakkasan company.

Echo Company (Recon) occupied abandoned FSB Scotch on the opposite end of the ridgeline. That evening the NVA attempted to overrun Scotch but Recon fought them off, killing at least seven NVA. Recon had only three slightly wounded.[195] According to army gossip, the total enemy dead was 25.[196] It should be noted that 101st Airborne Division policy prohibited reporting kills without bodies as evidence. The NVA only abandoned their dead when it was not possible to remove them. Hence enemy casualties certainly often were much higher than reported. Grunts frequently encountered heavy blood trails and/or body organs that were evidence of fatal wounds. However, these evident dead were not included in the official body count.

---

193. See Steve Vogel, *Through the Perilous Fight* (New York, New York: Random House Trade Publications, 2014), pp. 337-338, 348-350.

194. Nolan, *Into Laos*, p. 20.

195. Nolan, *Into Laos*, pp. 181-82.

196. Personal account dated "5 Sep 71."

After the battle one grunt from Echo Company found a quiet spot on that hill and cursed God out loud, "I swore my alligence [*sic.*] to Satan and spit as high as I could into the heavens. At that moment a wind came and caught that spit in mid-air [*sic.*]. My jaw dropped as that spit came back and hit me smack in the forehead."[197]

The incident "spooked" the grunt. Later in life he reversed his allegiance and became a devout Christian.

I received a letter from Mother in which she described watching a television news report that mentioned Firebase Scotch.[198] She hoped I was not near it. I found a spot where I could sit and not see Scotch — I deliberately faced away from it because Scotch was plainly visible from Purple Heart Hill. I don't remember the exact words I wrote, but I indicated that I could not see Scotch at the time. However, I did write a letter to Dale Cheney, a friend from high school, telling him where I was and what was happening...just in case something did happen to me.

After nearly one week on Purple Heart Hill, Charlie Company was relieved by Delta Company. The helicopters flying Delta onto Purple Heart Hill dropped Charlie into the valley below that blood soaked ridge. The two Rakkasan companies wasted no time in the exchange. Delta unloaded rapidly and Charlie boarded even faster. As the choppers lifted off we could see explosions near the base of the infamous Rock Pile. The sight of exploding artillery shells was all too familiar to the grunts in Charlie Company. The formation headed directly towards the puffs of gray smoke. *S***! That's our LZ.* During the time we occupied Purple Heart Hill, more than once we witnessed

---

197. In a post by William Morgan on the *VietnamWarHistoryOrg*. Facebook web site. Accessed February 8, 2016.

198. This may have been the CBS television report in which Lt. John Dewing was interviewed (see Nolan, *Into Laos*, 174).

green and red tracers in that same area. Green tracers belonged to the NVA and there had been "beaucoup"[199] green tracers in that valley. To our relief, the LZ was cold. Our mission was support for Alpha Company. On Thursday, February 25, Alpha Company destroyed a large bunker complex at the base of the ridge on which Purple Heart Hill and Scotch were located. They captured a major arms cache that included one Chinese 122mm rocket launcher along with its rockets (these may have participated in the bombardment of Purple Heart Hill on February 21). The size of the cache suggested a heavy NVA force nearby. Charlie's role was to augment Alpha's firepower and expand its effective search area.

As Alpha Company's operation unfolded, Charlie Company took some ineffective small arms fire, but encountered no serious enemy opposition. On the other hand, while tramping through the elephant grass we discovered artifacts from the terrible Marine Corps battles earlier in the war. Rusty M-14 magazines, oxidized 7.62mm brass casings, rotting web gear, and other debris lay scattered in various sites.

A few days later I received good news. The CO was sending me to China Beach for the in-country R&R that Colonel Sutton promised me for discovering the NVA ambush near Purple Heart Hill. I caught a ride back to Camp Evans on a resupply helicopter early the following morning. We flew at a high altitude on the way to Evans. I was sitting on the floor looking down at the jungle below when I spotted a bright green light in the jungle. Then it was floating up toward me. For a split second I wondered what it was. By then it passed me and went whizzing out of sight. Even as it registered that the light was an NVA tracer round, additional green lights followed its threatening path. *The d*** Dinks are shooting at me!* The Huey pilot

---

199. American slang for "many."

took evasive moves. In an instant he rolled the Huey like it was an Air Force fighter. For a moment I was horizontal looking straight down at the earth below me, parallel with the ground, held in the bird by centrifugal force alone. The entire episode happened so quickly that I did not have time to be afraid. The pilot straightened our flight course and we landed at Camp Evans without further incident. I picked up my travel orders from the company orderly room and headed for China Beach.

China Beach was a US military in-country R&R center. It was located on the ocean side of Tien Sha Peninsula near Da Nang. I don't recall much about China Beach, What I saw didn't resemble anything in the later television series named after the location.[200] And I didn't see anyone that looked like Dana Delany. While visiting the R&R Center, I wore civilian clothing. Activities included swimming in the South China Sea, reading paperback Western novels, and eating hot food at every meal. At the time, all that really mattered was that a grunt could take it easy because no one shot at you in the R&R center. With travel time, the three-day R&R kept me out of the bush about four days.

During the fighting around the Rockpile, Delta Company had their Claymore mines stolen by the NVA. The dinks crept up and cut the wires during the night.[201] As a precaution against similar incidents in the future, someone developed a devious countermeasure. Once the location for his Claymore had been determined, a grunt scratched out a small hole in the dirt. He placed a hand grenade in the hole with the safety lever on top. After setting the Claymore on top of the safety lever, he pulled the grenade's pin. The Claymore held

---

200. *China Beach*, created by William Broyles, Jr. and John Secret Young, ran from 1988 to 1991 on ABC television.
201. Nolan, *Into Laos*, p. 177.

the lever in place. Then the grunt covered the grenade with dirt. If the dinks tampered with the Claymore, the grenade exploded. Because of the risk involved, booby trapped Claymores only could be removed by the individual who set them up.

During Dewey Canyon II, the mountains near the Laotian border periodically were soaked with cold northern rains. Although the temperature rarely fell much below 50º F,[202] the wet conditions and wearing tropical clothing combined so that it felt extremely cold after dark. Occasionally grunts even could see the vapor of their breath. No matter what the thermometer might register, grunts got cold, bone chilling cold, when wet. On particularly cold nights, veteran soldiers employed a pragmatic countermeasure used by their forefathers in other wars. Two men lay down side by side on top of a poncho and wrapped up together with both of their poncho liners—a lightweight nylon quilted blanket. The two layers of cover retained the body heat that the soldiers generated, keeping them warm at night. One cold night a replacement spending his first night in the bush refused to participate in such "deviant" sleeping arrangements. The cold kept him up shivering all night. The next evening he was soliciting a spooning partner.

---

202. Website of Vietnamese Embassy in the United Kingdom (http://www.vietnamembassy.org.uk/climate.html). Accessed August 2, 2016.

DEWEY CANYON II/ LAM SON 719 (PART 2)

March 1971: The Mission's Objective Discarded

SAPPER ATTACK

The ARVN incursion into Laos in Lam Son 719 received priority in aviation commitments. Helicopter losses were extremely heavy. During 45 days, 82 helicopters were destroyed, or 14% of the choppers committed to the operation; 600 (68%) suffered combat damage.[203] The sight of high flying Chinooks with a damaged chopper in a sling underneath it was becoming too common.[204] The high losses diminished helicopter availability to resupply the American grunts fighting in the hills around Khe Sanh and the Rockpile. Moreover, weather created low ceilings and reduced visibility during 54% of the

---

203. Cosmas, MACV: The Joint Command in the Years of Withdrawal, 1968-1973, 335; Dunstan, Vietnam Choppers, pp. 42-43.

204. For a description of aircraft recovery by CH-47 Chinook helicopters, see Dunstan, *Vietnam Choppers*, pp. 81-83.

days of the operation.[205] When equipment losses were combined with frequent bad weather, the grunts experienced critical shortages. Normally an infantry company in the field expected to be resupplied with food every three to five days. During Dewey Canyon II going two or more days without any food was not uncommon. Consequently, grunts suffered long humps through the jungle and elephant grass on empty stomachs.

Aggressive action by the 4th Battalion of 3rd Infantry Regiment, 23rd Infantry Division (AMERICAL) and the 3rd Battalion of the 187th Infantry Regiment (Airmobile), 101st Airborne Division (Airmobile) in late February had led to intensive fighting near the infamous Rockpile. By the end of the month, enemy activity in the area dropped significantly. At the same time, operational control of the battalion reverted back to the 101st Airborne Division.

Life remained relatively mundane for the next week or so. Lieutenant Harris received word that his wife gave birth to their first child, a 5 pound 12 ounce baby boy name John Wayne Harris. I received a care package from home that included some of Mother's homemade fudge. There were a few minor firefights and several caches of NVA supplies destroyed, but no major contact with the NVA. Enemy forces had either been withdrawn in order to reinforce their units fighting the ARVN incursion in Laos or they had been rendered ineffective by combat with the Americans. Elsewhere, the campaign was far less successful. An NVA counterattack was pushing ARVN troops out of Laos. Khe Sanh was under constant shelling.

In mid-March, Charlie Company was pulled out of the AO around the Rockpile for redeployment northwest of Khe Sanh. Sufficient Huey choppers were not available both to extract Charlie Company

---

205. 101st Airborne Division (Airmobile) Final Report, Airmobile Operations in Support of Operation Lamson 719 8 February - 6 April 1971, IV-52.

and support the ARVN withdrawal out of Laos.[206] Consequently, we humped out of the mountains to link up with a tank company from the 1st Battalion, 77th Armor of the 5th Infantry Division (Mechanized). The unit was equipped with M-48A3 Patton medium tanks. The M-48 was a rugged battle tank equipped with the M-41 90mm gun.[207] The vehicles were sitting in a perimeter on top a hill near the highway when Charlie Company burst out of the elephant grass. Exhausted grunts climbed up onto the back of the M-48s. After the entire company piled on them, the tanks eased down the hillside and onto the highway. Tank engines reverberated underneath weary grunts as the armor column rumbled down the road toward Camp Carroll at 20-25 miles per hour.[208] The scene was reminiscent of documentaries on World War II. Along the way we passed an ARVN truck convoy going the opposite way. Someone tossed a green smoke grenade into the empty truck bed of one vehicle. The subsequent laughter of the grunts was drowned out by the growl of the tanks' engines.

The armored convoy reached Camp Carroll without incident. Camp Carroll—frequently referred to as Vandegrift or Vandergrift by Charlie Company's grunts and others in the campaign—for a night of rest.[209] Camp Carroll was located approximately eight

---

206. Cosmas, MACV: The Joint Command in the Years of Withdrawal, 1968-1973, 335.

207. D. Eshel, ed., *M-48/60 Patton Main Battle Tank*, War Data (Hod Hasharon: Eshel Dramit, 1979), p. 10.

208. Kevin Owens remembered the destination as Khe Sanh (in a telephone conversation on January 10, 2018). In examining my photographs, Charlie Company had been at Khe Sanh prior to Purple Heart Hill and at Carroll after Purple Heart Hill. Moreover, by late February, early March, Khe Sanh was taking heavy rocket and mortar incoming daily. Therefore, I concluded this move was to Camp Carroll.

209. Vandegrift Combat Base was dismantled in October 1969 and its materials

kilometers east of the Rock Pile.[210] At Camp Carroll, Charlie Company disassembled weapons and gave them a thorough cleaning and careful maintenance. We were fed hot chow and relished a few hours of relaxation. Temporary field hooches were erected in an old sunken roadbed. Only two men were assigned guard duty for the entire company at any given time that evening. A majority of the company enjoyed a full night of uninterrupted sleep for once. The next day we were sent back into the bush.

After spending the night in the relative safety of Carroll, Charlie Company made another CA. The entire 3/187th Infantry was relocated northwest of Khe Sanh. Charlie Company was inserted along the Laotian border. The NVA counterattack against the ARVN in Laos was now pushing into Vietnam. Our mission was to find and destroy NVA operating against the Khe Sanh Combat Base. Consequently, the battalion's grunts humped through the hills searching for the enemy. The seemingly aimless wandering made little sense to the grunts. We were hot and exhausted, but the close proximity of significant NVA units excluded negligence. So we sent out RIFs every morning. Then we moved to a new location before dark and dug in for the night. After sunset, one in two men commonly stood watch.

During Dewey Canyon II, Charlie Company made a CA about every five to seven days. On this occasion, the LZ was a small knoll covered in low elephant grass and undersized bushes. The size of the hilltop limited the CA to a single chopper at a time. A number of abandoned NVA bunkers were hidden in the vegetation. Off to one side a shallow saddle led to another bald peak. The whole site exhibited evidence of previous destruction, either by forest fire or Agent

---

used in the construction of Camp Carroll. Kelley, *Where We Were in Vietnam*, p. 5-532. However Vandegrift occurs in numerous official records of the campaign.
210. Kelley, *Where We Were in Vietnam*, p. 5-93.

Orange.[211] Numerous dead tree trunks towered out of the ground foliage. This second hill was higher but less suitable for an LZ because of these wooden shafts. On the lower hill where we landed, a solitary trunk stood at the edge of the LZ.

Several birds already had deposited their grunts when I landed. I bent down and raced downhill toward the growing perimeter. I reached my assigned spot and dropped into a prone fighting position facing away from the LZ. Seconds later, I heard a loud thud behind me. I turned my head and looked over my left shoulder. A brand new Huey had unloaded its passengers and was taking off when the tip of its rotary blade hit the dead tree. The helicopter hit the ground and burst into flames. A large fireball rose high into the sky. The crew all managed to escape but their chopper disintegrated in the scorching fire. I don't know how long the fire lasted, but it seemed like an eternity. Those few moments were scary. The fire was intense and the slick's M-60 ammunition was cooking off. Landing the rest of the company was impossible. Until the fire was extinguished, the troops on the ground were about half strength. Moreover the cooking machine gun rounds prevented spreading them out into a complete perimeter. Instead we hugged the ground and waited for the flames to die out. The fire ended almost as abruptly as it began. Only a pile of white ash, two machine guns, two rotary blades, the aircraft's engine, and a few other parts remained.

The danger from the burning chopper passed. The suspended CA resumed. The next bird into the LZ, having disgorged its grunts, picked up the stranded helicopter crew of the wrecked chopper. After the company reassembled on the ground, Charlie Company proceeded with its mission. First, we checked out a handful of bunkers

---

211. Agent Orange was a herbicide used in Vietnam for defoliation to deny the NVA the shelter of the jungle for concealment.

near the LZ. We found them empty. Lacking sufficient demolition to destroy them, the company proceeded to move to a higher hilltop adjacent to our location. As Kevin Owens, the point man, traversed the saddle and started ascending the designated hill, he met an NVA soldier. He quickly fired several shots, but the enemy soldier escaped.

Charlie Company advanced up the slope with caution. We reached the top and descended down the other side of the hill. Denser jungle vegetation covered the hillside. Numerous well camouflaged NVA bunkers emerged into sight under the leafy canopy. A grunt entered the nearest bunker and discovered bloody bandages. Apparently the NVA soldier that Kevin shot stopped there to bind up his wounds. Bunker complexes presented a multitude of hazards. They concealed well-fortified enemy forces. Empty bunkers often were booby-trapped. Therefore we pulled back to the LZ and called in an airstrike.

Yellow smoke marked our position as a pair of Navy A-4 Skyhawks thundered into sight. They circled briefly, then diving low over the target released their deadly cargo. As the two fighter aircraft pulled up, four napalm bombs tumbled earthward. They struck with vengeance. An orange fireball enveloped the hillside. Several secondary explosions reverberated from the flames.

While the two planes circled overhead, Charlie Company headed back toward the hilltop. By the time we reached its crest, the two aircraft had vanished over the eastern horizon. We plunged alone down into the abyss. A sticky honey-like fluid hung off several trees as we moved into the area of the airstrike. This gave way to black charred trees devoid of leafs. Flames still burned in some places. The temperature seemed abnormally elevated and breathing was difficult. Several bunkers had collapsed, but there was no sign of the NVA. The company did not linger. We returned to the hilltop and established a NDP. The following morning we moved down into the jungle contiguous to the site of the napalm strike and continued searching for the elusive enemy.

* * *

Colonel Sutton and Major Schornburg were killed in a helicopter crash on March 17. Schornburg had just returned to the battalion from a stateside leave. After being wounded on Purple Heart Hill, he had been sent home for convalescence. Sutton was replaced by Lieutenant Colonel James Robert Stevenson.

* * *

On March 20, Charlie Company was alerted as a reaction force to CA in support of a LRRP (Long Range Reconnaissance Patrol) from the 5th Mech. However, the situation changed and we did not make the CA.

On Saturday morning, March 22, first platoon moved out on a RIF. Around noon, we discovered and destroyed five small bunkers. A short time later, NVA soldiers wearing khaki uniforms were spotted moving through the elephant grass on a distant hilltop. The grass made it difficult to count them precisely, but we estimated the force numbered between 25 and 30 troops. They were out of range for the weapons we carried. The maximum effective range of the M-16 rifle is only 460 meters.[212] The company CO instructed the RIF to pursue the enemy in an attempt to engage and destroy them. We located their trail and tracked them for approximately thirty minutes or more through the elephant grass. Then their course veered sharply into a

---

212. Donald B. McLean, *AR-15, M-16 and M-16A1 5.56MM Rifles* (Cornville, Arizona: Desert Publications,1968), p. 23.

steep ravine covered in heavy jungle. At this point, the LT ordered us to end the pursuit and rejoin the company.

The senior NCOs and several veterans disagreed with his assessment and contended that the platoon should go into the jungle after the NVA. Dewey Canyon II had taken a toll on Charlie Company. Combat casualties, tropical infections, and other infirmities had ruthlessly depleted the ranks. Less than 50% of Charlie Company's authorized strength was in the field. This created a shortage of NCOs available for the command structure. Therefore as the ranking enlisted man, I temporarily was assigned the job of squad leader in first squad.

I agreed with the NCOs and other veterans. However, the LT was concerned that the NVA might set up an ambush and refused to grant permission to go. A brief debate ensued. Everyone recognized the very real danger that the pursued enemy force would turn and attack us on ground favorable to them. Nevertheless, many experienced grunts felt that an aggressive pursuit would not allow them sufficient time to prepare an adequate ambush. An old axiom in warfare says that the best defense is a strong offense. These veterans argued that our keen awareness of the risks would mitigate the danger. We could inflict more damage on the NVA with fewer casualties now than if we allowed the enemy unit time to attack us on their terms. The LT hesitated. Rather than enforce a direct order or yield to men with more combat experience, he issued no commands. Finally, he reported the situation (as he viewed it) to the CO and received authorization to discontinue the RIF. The decision would exact a heavy toll on the company later that night.

Charlie Company set up on a hill covered in short elephant grass.[213] The incline on three sides was relatively steep. On the fourth

---

213. Quang Tri Province, South Vietnam. UTM grid reference is XD726484 [DA2496 (RMN 64), CIL 53165, RoR 06161 Charlie 3/187, attack on NDP].

side however was a gentle slope that led down to a knoll covered in tall elephant grass and brush. Second platoon was assigned this sector of the perimeter. Recognizing the weakness of their sector, the grunts in second platoon set out all available trip flares, Claymore mines, and mechanical ambushes.

Although the company was extremely understrength, its grunts were confident. Repeated contact with the NVA in January and February had honed them into a deadly fighting force. The grunts dug in on the high ground of the best defensive position available in the nearby area. Freshly dug four-man foxholes ringed the hilltop. In a firefight, the two men on the outside corners of the hole provided enfilade fire in front of the adjacent holes on either side. The two middle men's fire covered the area directly in front of the position. This provided interlocking fire 360 degrees around the perimeter. Behind every hole was a sleeping position. This was a six feet long by four feet wide by two feet deep hole. Two men from each foxhole slept while the other two pulled guard duty. The men asleep were relatively safe from everything except a direct mortar round hit.

A company's survival in the bush required that an individual soldier perform various assignments. The most undesirable or menial tasks generally were assigned to the newest members in the company. In one sense this was a form of hazing just like occurs in a college fraternity. From a hardnosed perspective, these newer men were less valuable members of the unit. They could not contribute as much to the unit as the experienced veterans. Nor had they yet created any bond of camaraderie with other soldiers in the unit.

The practice also achieved pragmatic results. It integrated them into the unit and prepared them for the harshness of a grunt's existence. One of the common tasks assigned replacements was digging foxholes. The chore acclimatized their body to the physical demands of infantrymen in the field. It also shaped their mental conditioning to the necessary routine of digging a foxhole every night. Fresh replacements were not available most nights and so the veterans usually dug their own holes.

When the platoon received a couple of replacements that day, as the acting squad leader I put one into my hole where I could watch him closely. I would need to teach him the correct way to do things if he was going to assimilate fully into the platoon. I handed him an entrenching tool and told him to dig. The M1951 entrenching tool had a short wooden handle with folding shovel and pick heads. The replacement dug the foxhole slightly under four feet deep and began complaining.

I said emphatically, "Keep digging! If anything happens tonight, I am getting in the hole completely. If there's not room for you, you'll have to stay outside in the open." He dug out a few more inches, but it still was not adequate. I should have completed the hole, but I did not. Perhaps the rain and wind that evening influenced my decision. Perhaps it was due to being raw in the role of squad leader. Whatever the reasons, despite the known closeness of the NVA, I allowed the hole to remain too shallow. I would soon have reason to regret my decision.

Dave Carver and Bernard "Jenks" Jenkins were the other grunts in our hole. Dave was a slender white guy with an All American face. At first glance, he appeared much too young to be a soldier. On the other hand, Jenks was a brawny black kid from "Nawlins" (New Orleans, Louisiana). A mischievous smile always framed a set of slightly yellowed teeth against his chocolate brown face. Both were veterans, but Dave had more experience in the bush. I assigned the first guard watch to the replacement and Dave. I could trust Dave to keep an eye on the new guy. I would take the second watch so that Jenks and I would be on guard in the hours when a grunt is most likely to fall asleep.

At sunset I pulled off my boots and set them on the ground at the foot of my sleeping position. Jenks already was stretched out on his poncho. I barely could see him in the dark hole. I guess I was silhouetted against the sky. Jenks spoke softly, "Good night, Zero."

"Good night, Jenks. I'll see you in a couple hours."

"Yep."

Because of the earlier NVA sighting, everyone was skittish that night. Orders specified that half of the grunts on the hill would be awake at all times while the other half slept. Reality meant sleeping would be extremely difficult for everyone. An intermittent rain was falling. I knew I would have only a slim prospect for slumber. I lay down, rolled up in my poncho and liner, and tried to catch some sleep before my turn at guard duty. Eventually the physical exhaustion from the day's activities overtook me and I settled into a light doze.

I was awakened abruptly by small arms fire and explosions. Their close proximity startled me, but I instantly recognized the commotion was coming from second platoon's sector on the far side of the perimeter. I grabbed my M-16 and helmet and dove into the foxhole. My sock feet sunk into soft mud at the bottom of the cavity. *D*** I forgot my boots!* Glancing back just as another explosion momentarily illuminated the hilltop, I saw my boots just where I left them. The new guy looked at me. He was bewildered and frightened. I glanced at the other two guys in the foxhole. Dave and Jenks kept glancing back towards second platoon's position instead of peering in the blackness to their front. I immediately instructed Dave and Jenks to watch the area to our front. I quickly surveyed my surroundings, but in the dark could see nothing. Momentarily I stared in the direction from which I had heard explosions. They had come from second platoon's sector, but I could not determine their nature or cause. An eerie silence momentarily shrouded the entire hill. *What the h*** is going on? Too many explosions for a spooked grunt or an animal in a mechanical* *ambush…what the h*** happened?* Whatever had happened, it was not good. I realized our foxhole had to be deeper. Therefore I slid out of the hole and directed the cherry to dig and improve our fighting position. I knew he had not yet developed the stamina necessary for such physical labor. But I needed Alan and Jenks' veteran eyes peering into the darkness. They would detect any enemy probe long before the cherry could.

A scout dog had been assigned to the company that afternoon. The handler had dug a sleeping position for himself and the dog adjacent to my sleeping hole. I crawled back to his position and invited him to join us in the foxhole. I also snatched my boots. When I returned to the foxhole, the replacement was grumbling again.

"My back hurts!"

"Dig!" I screamed.

"This is deep enough," he said.

"No! It's not! Dig."

"It's over now."

"The h*** it is! Dig it deeper!"

"I can't do anymore!"

"D*** it! I said, 'Dig,' and I f***ing mean dig. I ain't gonna tell you again. If this f***ing hole isn't deep enough, I'm gonna throw your scrawny a** out of it and let Charlie blow you away. Now dig."

He kept digging. The replacement later became a good soldier, but that night he was a cherry. His body was unaccustomed to the rigors of life in the field and his mind had not developed an instinct for survival.

A short time later, more explosions rocked the perimeter. Again they came from second platoon's sector. This time we clearly could see satchel charges exploding in second platoon's foxholes. Their flashes also illuminated forms moving inside our perimeter. The flash was too brief to identify the forms. Someone popped a handheld flare.

A yellow tail marked the flare's course as it climbed skyward. It exploded and a soft glow illuminated the scene as the flare floated down under its tiny parachute. The light from the flare was sufficient to reveal khaki figures inside the perimeter. Chaos ensued. American grunts were firing M-16s and M-60s in every direction. AK-47 fire was equally chaotic. The NVA had struck second platoon and overrun it. The integrity of the company's perimeter collapsed. Suddenly the occupants of each surviving foxhole adopted an independent perimeter of 360 degrees around itself. The attack penetrated all the way to the company command post in the center of the perimeter before being stopped.

By now the replacement's shovel was on steroids. He was removing dirt like a front loader. Clods of dirt were flying and the floor was sinking. Soon I slid back into the safety of the hole. The cramped space did not deter the motivated cherry with his shovel. He continued to excavate our cavity in the earth. By dawn the hole was too deep. You barely could see out of it and it held five G.I.s and one large German Shepherd dog!

The entire perimeter was now blazing away in every direction. Red tracers zipped across the inside of the perimeter as startled grunts fired at the running NVA. As the flare's light faded away, the action shifted from inside the perimeter to outside. Any sound, real or imagined, brought an immediate response from jittery grunts. Thumbs flipped selector levers to fully automatic. Nervous fingers squeezed triggers, haphazardly spraying the darkness with lethal rounds. Hurriedly more experienced veterans seized control of the situation and ended this indiscriminate firing. In the dark the flash from a rifle shot revealed the shooter's location. Thereafter grenades were the primary weapon of choice. They did not disclose one's position in the darkness and were more effective at killing unseen enemies. The slightest sound provoked jittery grunts to pull the safety pin from a grenade and hurl it into the darkness. An ephemeral flash of light and a dull boom followed. Then silence...until someone else heard another noise. The flashes and explosions reverberated all night. Periodically the flash and boom were inside the former perimeter as the NVA probed our defenses. By morning, grunts had all but depleted the company's supply of these spherical bombs.

The sight of NVA inside the perimeter meant Charlie Company's ability to function as a unit no longer existed. Each foxhole now considered itself isolated and its field of fire extended completely around it. A long night of terror had commenced. No one knew the extent of damage suffered by the company. You only could be certain that the guy in your hole was alive. You might have glanced at the foxholes closest to yours during the flash of an explosion. Its occupants were alive at that moment, but were they still alive? The firing indicated others were alive too, but once the shooting died away even that was uncertain. The color of the tracers no longer provided clues to the identity of the combatant. The grenade duel raised as many questions as it answered. The few working radios on the hill maintained limited communication, both with each other and with the battalion. But the majority of grunts in the foxholes had no radio. They were cut off from each other physically and psychologically. *How many survivors remained?* Lacking accurate information about the situation, wild speculation filled the void. Most grunts struggled with a feeling that only a few grunts had survived the night's attack. Fear that another assault would be sufficient to wipe out the rest of the company hung heavy over everyone.

In that moment of deepest despair, the first welcome sign of hope divulged itself. An artillery shell burst high in the night sky. A large flare floated out of its smoke, illuminating the entire hilltop. Then a second round exploded just as the light of the first flickered and died away. Then a third and a fourth in succession. Their glow assured us that we could see any attacker before he reached our foxhole. The existence of these artillery illumination rounds also indicated that someone was in communication with the rear. The supporting battery continued firing illumination rounds. A cat and mouse battle raged all night. American grunts defended their hole resolutely by tossing grenades at every sound. The NVA sappers attempted to get closer as the flares dimmed.

Flares floated earthward on miniature parachutes. They swung back and forth as they descended, leaving a wavy trail of smoke behind them. Their ghostly light illuminated this smoke trail so that it was plainly visible against the charcoal colored night sky. As soon as a flare started to die away, the NVA again crept up and tossed grenades at the perimeter. Any sound outside the perimeter instigated the grunts to hurl grenades and fire M-203s into the darkness. Periodically the artillery shifted to high explosive rounds, temporarily delivering a virtual wall of detonating explosions beyond one sector of Charlie Company's perimeter.

I eyeballed our foxhole. Dave and Jenks alertly watched front and rear. Now knowing I was not alone on that hilltop, I crept out of my foxhole once again and low-crawled to the foxhole on the right side of my position. Once I made contact, I turned and slithered over to the hole on the left side. It was risky but necessary. If I exposed myself while a flare was in the air, I gave the NVA a clear target. If I moved in the dark, my buddies might shoot me. I was not the only grunt making contact with flanking holes. Other grunts around the perimeter were doing the same. Thus we started to establish a contiguous defensive line. As I scurried back into my foxhole one of the company RTOs scooted up to me. Battalion ordered Charlie Company to set fire to the grass around the NDP to enhance our field of fire and to better mark our location in the developing fog for support aircraft. All around the company perimeter others also were making contact with adjacent foxholes. By dawn a partial perimeter had been reestablished. Second platoon's sector was gone, however. At least four grunts were dead and six others wounded. Daylight brought the discovery of another dead GI. As the warm sun rose higher in the morning sky, the embattled grunts took assessment of their situation. Only three dink bodies remained. All three were less than a foot from the Company CP's foxhole in the center of the NDP.

At one point that night a "Spooky" (also known as "Puff the Magic Dragon") was on station dropping flares and firing at the ground below. Spooky gunships were Air Force cargo planes equipped with multiple 7.62mm miniguns.[214] I don't know if this one was an AC-47[215] or the later AC-130.[216] Either plane was highly effective. Their guns had a rate of fire of 6,000 rounds per minute.[217] Every fifth round was a tracer. When shooting, the tracers looked like a red-orange rope curving down out of the aircraft and tying it to the ground. We did not know what Spooky was shooting at, or even if it was in support of us, but that unbroken line of light emitted by its tracers brought comfort to the grunts on the ground that night.

Documents on the enemy corpses indicated they belonged to a Sapper unit on its way to hit Khe Sanh. These elite units were trained to maneuver through barbed wire and the substantial defenses of

---

214. Antiquated C-47 were first employed. These later were supplemented with AC-119 and C-130s.

215. The AC-47 was the military version of the Douglas Aircraft DC-3, a World War II era cargo plane. The Vietnam era gunship mounted 3 miniguns in the rear of its fuselage.

216. The AC-130 was a Lockheed C-130 cargo plane. Rotary 20mm cannons were added to the armament of the AC-130 in 1969.

217. Bernard C. Nalty, "The air war on the Laotian supply routes," *The Vietnam War*, ed. Ray Bonds, p. 169.

heavily defended permanent sites. The puny trip flares and other early warning devices set out by an infantry company in the field were ineffective against such well-trained professionals.

The sun rising slowly in the east ended the night of terror. The sporadic noise of explosions and gunfire died away. With daylight, Dust-Off flights medevac'd the wounded. Their Hueys, emblazoned with the iconic white square containing a red cross, alternated landing with resupply birds rushing ammunition to replace what had been expended in the night. Not surprisingly, casualties were higher than first estimated. In addition to the five dead, 12 wounded grunts required evacuation. The CP only popped smoke for the first chopper. The others followed in such rapid succession as to make smoke for them superfluous. Before the first Dust-Off lifted off the LZ, the battalion commander's C&C bird arrived and circled overhead. As the last Dust-Off approached, another UH-1H Huey helicopter arrived above the hill. It was a resupply bird with ammunition and C–Rations. It too started to circle, waiting for an opportunity to land. As quickly as it landed a detail of grunts swiftly unloaded the supplies. Boxes and crates were pulled out rapidly and flung onto the ground haphazardly. The bodies of PFC Reginald J. Abernethy, SP4 William H. Fells, PFC William S. Glenn, PFC John W. McLemore Jr., and PFC Roger W. Stahl were loaded on the bird.

Abernethy, age 22, was from Maiden, North Carolina. He had been in country less than three months. Fells, age 20, was from Washington, D.C. He had arrived in Vietnam on February 14. Glenn, age 22, was married and came from Akron, Ohio. He too had been in Vietnam less than three months. McLemore, age 24, was from Fresno, California. He was married and had been in Vietnam 32 days. Stahl, age 21, was from Somerset, Pennsylvania. He had over five months in country and was promoted posthumously to Corporal. Their names are inscribed on panel 04W of the Vietnam Wall in Washington D.C.

As soon as the resupply bird took off, Colonel Stevenson's C&C landed. The night also had been long and anxious in the battalion rear. Colonel Stevenson spent the night monitoring the situation in the TOC.[218] First Sergeant Herrington also listened nervously. First Lieutenant Steve Metcalf, second platoon's former platoon leader, joined the worried party at some point. As the severity of the episode became more and more apparent, Stevenson turned to Metcalf and asked rhetorically, "Do you want to take command of that company?"[219]

"Yes sir," responded Metcalf, recognizing the question was an order, not a request.

Now as the chopper's skids touched the ground Stevenson bounded out and raced toward the company CP. Metcalf tactfully tailed the battalion commander by a few yards. He remained a discreet distance as Stevenson spoke to the company commander. The captain was relieved on the spot.

An infantry company commander has to be one of the most demanding jobs in the world. He is responsible for everything pertaining to that company. Its survival is dependent upon him. He must make certain that his men are properly supplied with ammunition, clothing, equipment, and food. A good RTO assumes a significant portion of the administrative tasks, but even here the CO cannot avoid accountability. He selects the man to be his RTO and conveys to him the expectations for the job. The company commander determines what the men will do and when they will do it. Insuring the company acts as a cohesive fighting unit is his responsibility. At the

---

218. Tactical Operations Center.

219. In a telephone conversation with Metcalf, April 21, 2017, Metcalf said Sutton tapped him for command of Charlie Company. However Sutton was killed on March 17.

same time, he has the entire chain of command above him coercing him to accomplish the mission assigned to his company. His superiors may not have the slightest ideas what the true situation on the ground actually is. If the mission fails or disaster strikes, blame will fall squarely on his shoulders.

Circumstances on a battlefield are fluid and change rapidly. Therefore an infantry company commander must be astute at assessing any situation and fast in formulating the best action to meet its demands. Herein lays the breaking point for many commanders. The enemy plays a major role in determining the battlefield situation. And the enemy frequently behaves in an unanticipated manner, creating surprising scenarios. Under the stress produced by the new circumstances, company commanders sometimes collapse. Some slip into paralysis. Others become delusional. So they fail. Others are classified failures because in the circumstances success was impossible. That night the enemy destroyed second platoon as an effective fighting force. The company commander could do nothing to prevent it, but he was the commanding officer. Therefore the company commander was held accountable.

Around the perimeter, grunts distributed the ammunition and supplies dropped by the resupply birds. We partially packed rucksacks—wet ponchos and poncho liners were spread out to dry in the morning sun—and started policing the area for whatever the enemy had left behind. At the same time, we constantly glanced at the three officers standing in the center of the NDP. Stevenson and the CO faced each other only about two feet apart. The CO hung his head as the Colonel spoke. Metcalf waited about eight feet away from the two. He stood almost in a parade "at ease" posture, holding his own hands behind his back. An explanation was not necessary. All but the newest cherries realized Metcalf was replacing the previous commanding officer.

Many heard Stevenson yell, "Metcalf!"

Most saw Metcalf bolt over to the battalion commander. As the two officers spoke, the relieved company commander mopped over to the CP where he commenced to gather and pack his equipment. After a brief conversation, Stevenson and Metcalf made a short inspection of the NDP, observing the grunts primarily. Occasionally they stopped and asked a question or two. By the time they completed their assessment, the relieved CO headed to the colonel's C&C bird. Stevenson joined him inside the chopper. It quickly lifted up and flew away.

Metcalf immediately assumed command of Charlie Company. Metcalf had commanded second platoon during the Falls Operation. Grunts greeted his return as company commander frostily. He had been considered far too arrogant and too gung ho as a platoon leader. So now prophets of gloom foretold of our death and destruction. In time, Metcalf would negate most fears and earn respect as a capable combat commander. However, on that chilly March morning, Charlie Company's grunts viewed his assignment as a harbinger of a bloody future. They had survived the worst night of their lives, but under the new CO they feared more terror filled nights soon would be coming their way.

Nonetheless, First Sergeant Herrington began a relationship that revealed the grunt's prognostication of Metcalf was incorrect. Before leaving with Stevenson, Herrington approached Metcalf and said, "Come with me to the wire." The two professional soldiers stepped away from everyone else. The older NCO posed some procedural questions interspersed with a few insights.

Metcalf listened attentively and then asked, "How long have you been in the army, First Sergeant?"

"Twenty-two years, Sir."

"Continue, First Sergeant."

Herrington made a few additional comments to the new company commander.

Then Metcalf responded, "First Sergeant, you continue running this company the way that you think is best."

After that, cigarettes and toothpaste for Metcalf mysteriously appeared on the resupply bird just as he was getting short of those items. Herrington later confessed that he sent toothpaste when his own tube got low. As for the cigarettes, the company clerk also was a smoker. When he needed a new carton, Herrington added a carton to the CO's resupply.[220]

Metcalf's aggressive combat leadership was driven not by ambition, as was widely believed, but by determination to keep his men alive. He initially had been assigned to the Fourth Infantry Division and had seen combat with that unit. He had studied action in Vietnam and discovered units that ceased active patrolling and allowed discipline to deteriorate soon experienced high casualties. He believed a good leader kept the fighting ability of his command sharp.[221]

In his book *The Life of Johnny Reb* the noted historian Bell Irvin Wiley observed that many soldiers in the American Civil War were "appalled by their growing insensitiveness to human suffering and death." He added that many others had become so thoroughly hardened that they did not even think about the viciousness.[222] This callousness is universal to an infantrymen's experience. It enables him to survive the brutality of war. Later that morning, I surveyed second platoon's position. Lying in the dirt was a human brain. It

---

220. In a telephone conversation between Herrington and the author, April 28, 2017.

221. In a telephone conversation between Metcalf and the author, April 21, 2017.

222. Bell Irvin Wiley, *The Life of Johnny Reb: The Common Soldier of the Confederacy* (Garden City, New York: Doubleday and Company, 1971), p. 35.

had the appearance of having been removed undamaged by a medical student's dissection rather than by a violent act of combat. Since the skulls of the American dead remained intact, the organ obviously was that of a dead NVA soldier. All around was more unmistakable evidence that the previous night's battle had not been as lopsided as it seemed in the dark. Blood soaked paths made by dragging human bodies over the ground indicated enemy casualties exceeded the three dead NVA lying near the CP. Swarms of flies hovered over the debris of the battle. The dead NVA corpses and discarded bloody bandages scattered on the ground especially attracted the insects. Abandoned AK-47s also were covered with gore. After inspecting the sector, I sat on the ground nearby with a couple of buddies and ate lunch. The dead NVA and other bloody carnage were distinctly noticeable, yet the gruesome view did not adversely affect our appetite. We chewed our cold C-rations and joked about gorging ourselves on a McDonald's Big Mac® when we got back to the world.

THE BUNKER COMPLEX

Two days later, Charlie Company was placed on alert for a possible CA in support of Delta Company. However the CA became unnecessary and the mission was canceled. That night the NVA struck Charlie Company again. This time they were spotted moving into their assault positions. We opened up on them before they were ready. Consequently the NVA attack failed to reach our perimeter. They fired AK-47s and RPGs before being driven off. One grunt was slightly injured. The next morning we discovered several heavy blood trails.

The following day, third platoon was in point. Lieutenant Howie Cantley and two men were cutting trail when they uncovered an old

NVA bunker. The NVA constructed bunkers according to one of two designs. This bunker followed the first pattern.

A typical NVA bunker in the mountains adjacent to the coastal plain was a three feet by six feet hole approximately four feet deep. The dirt floor was angled slightly so that grenades would roll into a grenade sump. These sumps were similar to a golf hole, only deeper. They absorbed the shrapnel from the explosion, protecting the occupants from severe harm. Logs, six to eight inches in diameter, roofed the hole. The logs then were covered with a two to three feet deep layer of dirt. Commonly only about 18 inches to two feet of the roof was above ground level. In the fertile topical climate the layer of soil soon was masked with fresh vegetation, making it extremely difficult for us to detect until it was too late. A narrow opening with a shallow step was cut at each end of the bunker. The opening provided access into the bunker and served as a firing port during combat.

Further west in the mountains along the A Shau Valley and in the region around Khe Sanh, some bunkers were larger and constructed with an A-frame roof. A row of log columns ran down the center of the bunker to support a log ridge beam. Logs were laid with one end on the ridge beam and one end on the ground to form a roof. This roof had a pitch of approximately 30° on each side of the ridge beam. Dirt also was piled on top of the logs and so the vegetation also made them difficult to spot. These larger bunkers often were connected with tunnels or trenches and sometimes were built in conjunction with the smaller bunkers.

When the cutting party encountered the bunker, Cantley sent one man back to bring up the rest of the company. His departure left Cantley and one man alone on the perimeter of an abandoned NVA defensive position. As the two exposed trail cutters waited, NVA in an unseen bunker opened up on them. An RPG struck a tree and exploded. Metal shrapnel and wood debris rained down on the two exposed soldiers. The steady kak-kak kak-kak of AK-47s superseded that lone RPG. The enlisted grunt dropped to the ground and returned fire. Cantley dove away from him in order to enhance separation between the two impromptu firing positions. Unfortunately, he lunged head first into the abandoned bunker and became wedged upside down in the opening. His legs protruded up into the air. They were kicking furiously when the company's lead elements arrived. A couple of grunts grabbed his legs and pulled him free as they extended the company's front. By this time the firing had intensified dramatically.

First Sergeant Herrington raced forward to the sound of combat. The sight of a terrified sergeant curled up in the fetal position disgusted the old veteran. The frightened sergeant's men were aggressively fighting to defeat the enemy while he sought safety behind them. However, before Herrington could reprimand the cowardly conduct he saw Cantley. The lieutenant was lying unconscious in a puddle of blood alongside the freshly cut trail. Initially, Herrington supposed he was dead. He and the senior medic dragged Cantley away from the firefight. Blood covered his head but closer inspection revealed that the officer was still alive. About this time, Cantley regained consciousness. Quickly, the First Sergeant grabbed him and assisted him back to the rear. They jumped over a fallen tree and landed in an anthill. Moving on they found another log without ants and it provided cover from enemy fire. At that moment the emotion built up inside Cantley overwhelmed him and he cried uncontrollably. Nevertheless, he insisted the First Sergeant not let anyone know about his tears. He had no reason to be ashamed. No one would have thought any less of him for crying. Under the circumstances, most of us would have done likewise.

After the firefight, Cantley's helmet liner was lying relatively undamaged near the bunker. When the helmet was recovered it had a bullet hole just above the camouflage band in the front. A second hole was located in the same location on the back side of the helmet. When he placed it on his head, his appearance solicited expressions of astonishment. He looked as if he had been shot in the forehead, but he plainly was alive. Closer examination of the two headgear pieces revealed that the AK-47 round had penetrated the steel pot, but ricocheted off the fiberglass liner. The copper-jacketed bullet had then travelled along the path of least resistance over the liner until it encountered the folded rim of the metal cover. The rim deflected the round, but it still possessed sufficient velocity to pierce the steel pot again. Cantley was medevac'd and survived. Years later he contacted Herrington and thanked the NCO for saving his life that day.[223]

A day later fourth platoon engaged an NVA force of unknown size.[224] The two sides exchanged fire briefly and then the NVA disengaged. They disappeared before the rest of the company could come on line. There were no known casualties on either side.

Company strength had dropped to under a hundred men in the field before the decimation of second platoon. With its loss, troop strength in Charlie Company became perilously low. This led to an influx of replacements and some reorganization of the company. The strength of all platoons was increased. An NCO took over as squad leader of first squad and I returned to being a RTO.

In order to keep second platoon from becoming totally green, a few veterans from the other platoons were moved into second platoon. One of these veterans shot himself in the foot when he learned

---

223. The account of Cantley is based upon my personal memories and an account given to me by CSM Herrington in a telephone conversation on January 23, 2017.
224. Normally fourth platoon was a heavy weapons platoon (mortars, recoilless rifles, etc.). Loss of equipment and the need for more riflemen temporally reassigned the platoon to line duty.

he was changing platoons. We heard that he was later court-martialed, but I cannot verify the reliability of the report. He had been in an OP the night the sappers hit the company. From his forward position he witnessed the sappers slaughtering second platoon but was helpless to stop it. Consequently he blamed himself for their deaths. He was no coward—previously he had demonstrated extraordinary valor. However, his sense of guilt and the separation from his buddies in first platoon pushed him over the edge. To my knowledge, no one within our platoon ever disparaged him for reaching his breaking point.

Some replacements told us about their experience when coming into Charlie Company. The incident is telling. During their convoy to Camp Evans and during training at SERTS, they constantly watched Chinooks flying overhead with damaged Cobras or Huey Slicks suspended underneath. At SERTS they learned of heavy fighting around Khe Sanh and near the A Shau Valley. The morning they arrived at the company orderly room, they had to wait for the company first sergeant. When he finally came into the room, Herrington ignored them and walked straight to the company status board. The board was a chalk board on which the name of every member of each platoon was written in chalk along with his daily status. The first sergeant calmly erased the names of the men in second platoon who had been killed. Next he wrote beside the name of each wounded soldier the medical facilities where he was located. Then he turned and looked at the replacements. He was silent for a moment; then he asked unsympathetically, "Do you know what platoon you are going to?"

After returning to the platoon CP as a RTO, I set up a hooch with Lieutenant Harris each night. Some members of first platoon held Harris' decision not to pursue the NVA as partially responsible for the company being overrun. One night we were positioned on a steep slope. The position directly up the hill from us also had a PRC-25 radio. That night they called LT and asked him about the bursting radius of a fragmentation grenade. He quickly cited the technical data and then inquired their reason for the question. The voice on the radio boldly responded that they were concerned about shrapnel hitting them if a grenade accidently rolled down the hill into the LT's position and exploded. Their answer clearly was a veiled threat to frag the LT. A few days later he received a transfer to a job in the rear. The timing was coincidental, but fortuitous. Officers routinely transfer to new jobs in order to expand their military experience. B. J., the senior NCO, took over the platoon. Harris was a great guy and a good peacetime soldier but he was less than stellar as a combat leader. B. J. was an experienced veteran on his second tour in Vietnam. With him as platoon leader and the increased number of grunts in the platoon, first platoon's combat efficiency improved.

The company dispersed into platoon-sized RIFs in an effort to find more of the enemy. These patrols operated in relative proximity of each other so that a platoon could be quickly reinforced if necessary. Excellent coordination was essential to preventing a friendly fire incident. On this particular afternoon, we were moving through elephant grass approximately eight to ten feet high. Late in the afternoon, B. J. decided to send out a small patrol to investigate a nearby hill as a possible NDP. Unfortunately, first platoon had only one workable PRC-25 radio. Therefore maintaining communication with the company necessitated that the lone PRC-25 remain with the platoon. The patrol would have to go out without any communication. Unfortunately, it was my radio that did not work.

B. J. tapped two replacements and me to make the patrol. We dropped our rucksacks and went light. I took note that the platoon's point man had stopped under a lone tree that reached high into the sky. The tree furnished a visible landmark through the adjoining elephant grass. I took point and we pushed our way through the elephant grass into the jungle until we ran into a high-speed trail. The compacted dirt surface was three to four feet wide, typical of NVA footpaths in the area. Its trampled appearance suggested it still was in use. However it headed straight up towards the hilltop that B. J. assigned me to investigate. We warily checked it out. The path was relatively steep here. After 20 meters, it curved sharply to the left and back into the elephant grass. The ascent here was less pronounced, but clearly seemed to head towards our objective. Accordingly, we vigilantly followed it for about 75 meters. At that point the path veered sharply downhill. Consequently, we left the trail and continued to the top of the chosen knoll. Whereas the trail had been a slight incline, our new course was comparatively steep until it leveled out on the crown of the hill.

The hill showed evidence that an American NDP had been there previously, probably in 1969 or 1968. Otherwise the site looked acceptable for our NDP. So we started our descent down the slope. We retraced our steps until we reached the high-speed NVA trail. We paused there briefly. Seeing no signs of the enemy, we turned onto the pathway and continued our return to the safety of the platoon. I was in the lead and the downhill decline allowed for a brisk pace. I strode around the sharp curve we encountered at the beginning of our climb up the trail. Suddenly found myself face to face with someone hiking up the trail in dirty olive green fatigues. I could see a line of other soldiers behind him. Momentarily we both froze, startled by each other's presence. At the time everything appeared to be in slow motion, but our assessment of each other could not have lasted more than a split second.

The man looked uncannily like a guy in third platoon and I wondered what third platoon was doing so far from their reported location. As I pondered the potential tragic consequences of two American units meeting unexpectedly in the bush, we continued to stare at each other. Rapidly his true identity registered with me. I observed his belt buckle. It was a standard issue NVA flat rectangular buckle with a star engraved in its center, but the star had been painted red. Suddenly I could see him distinctly. *Pith helmet. AK-47. Bed roll over the shoulder. How could I be so blind and stupid!*

Simultaneously he must have recognized that I was an American soldier. He swung his AK-47 up and opened fire. Before his first round whizzed past me, I shifted my M-16 to my right hip and automatically pointed it at the NVA soldier. The situation was precarious. I had two inexperienced kids with me. An NVA column of unknown strength blocked our path back to safety. We were now cut off without any means of communicating with the platoon. For an instant I panicked. The selector lever on an M-16A1 is located on the left side, just above the pistol grip. My right thumb pushed down hard on the selector lever and I squeezed the trigger. The selector lever had three positions. It had been on "safety" when pressed down. A single click would have moved the lever to a vertical position and rendered the rifle into a semi-automatic mode. But I pushed so vigorously that the lever clicked twice. The weapon was in a fully automatic mode. It would continue firing until I released the trigger or emptied the magazine. As a veteran combat infantryman, I knew not to employ the "auto" position. The kick in automatic mode caused the lightweight weapon to elevate rapidly. By the third or fourth round you are shooting high over your target.

I dropped to a prone position as I fired. The last bullet left the bolt locked open. Reaching for another magazine, I realized my two comrades had moved up on my right side and were firing as well. I inserted a full magazine and then moved the selector lever back to the "semi" setting. I fired three or four more rounds and motioned to cease fire. The AK-47 fire had stopped.

My jumpy automatic burst apparently had achieved one positive result. The NVA were as startled as we were. Like us, the unexpected presence of enemy soldiers so close had frightened them. Perhaps they misjudged the size of our patrol or were just plain scared. Whatever the reason, they had disengaged. Our situation remained perilous however. The NVA may have broken off contact to await a larger force coming up behind them. Or they merely may have been doing the same thing we were, assessing the situation before making their next move. Whatsoever the NVA might be doing, we still hypothetically were cut off from our platoon. We needed to get back to them quickly if we were going to escape. Otherwise we likely would become an MIA statistic, but in reality either dead or a POW. There was no room for error and there was no time to waste.

I turned and scanned my surroundings. The tree where our platoon had halted was barely perceptible through the top of the elephant grass. I pointed to the tree and whispered instructions to my two companions, "Run to that tree and yell as loud as you can until you get there."

They both looked bewildered. One asked, "Yell what?"

"Your name. Anything," I replied. "Just let them know who you are."

I knew that our platoon undoubtedly had heard the firing. They would have recognized the distinctive sounds of both the M-16 and the AK-47. But, they had no idea what the outcome had been. Therefore they almost certainly would fire at any movement in the tall grass. The two cherries did not grasp my reasoning fully, but they trusted my experience enough to do as I said. They sprang up and took off. They ran like a herd of startled antelope and yelled like flock of wild geese.

Lying prone on the ground I methodically squeezed my rifle's trigger. My selector lever remained on "semi." The M-16 responded with a cadenced crack, spraying the jungle about twelve inches above the trail surface. I figured the low trajectory would hit anyone pursuing us. A leg wound was not fatal but it would end or at least slow pursuit. As soon as the last round was discharged, I pressed the magazine release button and removed the empty magazine. I pulled a full magazine out of the bandolier hung across my chest and pushed it into the empty weapon. After chambering a round, I checked my front for any sight of the enemy.

Nothing. Better be sure.

Holding the M-16 in place with my left hand, with my right hand I reached down and removed a hand grenade from my belt. I placed the M-16 softly on the ground. With my left hand I pulled the pin out. Holding the safety lever in place in my right thumb, with my left hand I retrieved my assault rifle. Quickly I rolled over onto my left side and tossed the grenade in the direction where we had last seen the NVA. Cling! The safety lever popped away from the spherical bomb. Bam! The grenade exploded, hurling shrapnel and dirt in every direction.

The instant the grenade exploded, I jumped up and followed my two rookies. After going about 25 meters, I dropped and repeated the process of firing and lobbing a grenade. Then I again jumped up and ran, yelling, "It's Zero!" (Zero was my nickname.) We got safely back to the platoon without further incident.

I reported to B. J. and told him about the acceptability of the proposed NDP and the encounter with NVA. The entire platoon returned with me to the site of the encounter. We dropped rucksacks and sent a RIF down the trail. The RIF found a heavy blood trail, but no NVA. Footprints suggested they were running when they fled. We obviously had hit one, but did not drop him. *Unacceptable! If you don't kill or capture the enemy, **he will come back** and kill you.* After following the trail for about 300 meters, the RIF returned.

The platoon made its way up the hill and dug in for the night. A mechanical ambush was set up on the trail.

***

A mechanical ambush was an ingenious American booby trap fabricated from Claymore mines. Claymores are electrically detonated. Normally a hand detonator called "the clacker" was used to propel an electrical charge through a pair of wires to the mine. In a mechanical ambush a radio battery was substituted for the clacker. One of the two wires was cut close to its connection on the Claymore. Metal clips used for shipping M-16 ammunition were attached to each end of the wire at this cut. The clips were lashed together with a rubber band. A plastic spoon was inserted between the clips to break the circuit. Then the wires were attached to the battery terminals. A tripwire was fastened to the spoon. The Claymore was aimed at the trip wire. When an NVA soldier caught the tripwire, it pulled the spoon out. The clips snapped together completing the electrical circuit, thereby detonating the mine. Often other Claymores were daisy-chained together with det cord, increasing the effectiveness of the ambush by enlarging the kill zone. Another effective variation was adding fragmentation grenades to the daisy-chain and hanging them in trees above the kill zone. Setting up mechanical ambushes around an NDP was standard procedure for infantry companies in the bush. Because of the diversity of variations possible, only the man who set out the ambush was permitted to disassemble one.

***

The brief contact on the trail kept everyone alert. The memory of those sappers destroying second platoon was still too fresh in most minds. The platoon stayed on 50 percent alert that night, but nothing happened. I was physically and mentally exhausted. Consequently, I slept well when not on guard duty. However, my failure to bring down the lead NVA soldier caused countless nightmares later. I knew that for an instant I had panicked when I pushed my selector lever to fully automatic mode. A very different fate for me was highly plausible if the NVA had responded more methodically. I easily could have been killed or taken prisoner. I never again hit my selector lever so forcefully.

WITHDRAWAL FROM KHE SANH

The old Marine combat base at Khe Sanh was reopened at the beginning of Dewey Canyon II. By then the mere name Khe Sanh held a mystic spell over American forces that generated fear and apprehension. Special Forces first constructed an airfield on the plateau near the remote village of Khe Sanh in 1962. The site was only six kilometers east of the Laotian border and 24 kilometers south of the DMZ. The Marines set up a permanent base camp around the airfield in 1967. It was rectangular shaped, approximately one mile long and one half mile wide.[225]

The combat base had earned notoriety in the 1968 Tet offensive. The NVA launched an attack against it on January 20 of that year. For approximately 75 days the Marines were cutoff and surrounded in the base. Heavy fighting continued into July of that same year. Shortly thereafter the base was abandoned.

The Khe Sanh plateau effectively was surrounded by higher hills on all sides. Denying the NVA possession of these hills was essential to keeping Khe Sanh as a viable base camp. During the latter phase of Dewey Canyon II, Charlie Company was assigned a role in securing

---

225. John J. Tolson, *Airmobility in Vietnam: Helicopter Warfare in Southeast Asia* (New York, New York: Arno Press, 1981), p. 165.

these ridges. Despite the close proximity of the Combat Base, resupply was difficult. Heavy fog frequently enveloped the hills, making it problematic for the helicopters to fly at certain hours, especially in the early morning. At other times the fog buried the plateau. On those occasions, the hilltops were exposed like the tip of an iceberg in the ocean. Visibility on these exposed peaks was virtually unlimited. Nevertheless, the thick fog shrouding the combat base kept the helicopters grounded, unable to fly except in a dire emergency…and a grunt's empty stomach apparently was not a dire emergency.

On March 15, 1971, the Rakkasans' battalion headquarters was established at Khe Sanh. Periodically thereafter, Charlie Company was rotated into the base to defend the combat base's perimeter. While Charlie Company manned the bunkers on the green line around Khe Sanh, I observed an Air Force C-130 Hercules drop supplies into the base. It taxied down the runway, but never stopped. As the airplane was landing, the NVA dropped a few mortar rounds in its path. Its crew shoved pallets out the rear cargo door onto the tarmac. The plane was airborne long before it reached the end of the runway. Once its cargo was jettisoned, the C-130 rose sharply and headed for the safety of its base. The incident marked the dramatic shift that occurred in March. In Laos the NVA counterattack was sweeping the ARVN incursion back into Vietnam and escalating NVA attacks against American forces in and around Khe Sanh.

Before March 15, one or two C-130s often sat parked along the runway at Khe Sanh. After that date, the NVA fired artillery, mortars, and 122mm rockets at the base constantly. Up to 200 rounds hit it daily. By March 19, the NVA were firing approximately 20 rounds per hour continuously.[226]

One evening Charlie Company set up an NDP near the Laotian

---

226. Nolan, *Into Laos*, pp. 325-26.

border. Our reported map coordinates alleged that we were approximately one kilometer inside South Vietnam and one kilometer south of the DMZ. Sometime after dark in the distance west of our NDP, we heard the low squealing rumble and clanging metallic sounds. At first, the noise was indistinct and its identity dubious. The first claims that we were hearing enemy tanks at the outset were received with skepticism and teasing. As the noise became more discernable, every ear strained to identify its nature. Soon the sounds were undeniable. Armored vehicles were bearing down on our position. Neither the United States nor South Vietnam had any vehicles anywhere near our location. Only the NVA could be operating tanks in the immediate vicinity. North Vietnam had committed PT-76 light tanks along with T-34 and T-54 main battle tanks to their counterattack in Laos. That counterattack was spilling over into South Vietnam. The United States Congress may have prohibited American ground forces from entering Laos, but the National Assembly of the Democratic Republic of Vietnam (North Vietnam) had not prohibited its troops from leaving Laos and entering South Vietnam.

When the realization dawned on everyone that a potential encounter with NVA tanks was credible, all levity ceased. Sound carries at night. So I am uncertain as to the distance that separated us from its source. Nevertheless, grunts began preparation for an encounter. Officers and NCOs inventoried the number of M-72 LAWs available within the NDP. These handheld rockets were our only anti-tank weapons and we only had three. We could not repel an armor attack if it contained more than three tanks. And even then, every LAW fired must destroy its target. If either of these criteria wasn't met, we likely would be annihilated. The company CP duly reported the situation to battalion. Battalion promised to send us support. So now the question became, "Who would reach us first? The promised support or the NVA tanks?"

Soon a battalion radio transmission disclosed an Arc Light strike on its way to the Ho Chi Minh Trail had been diverted to target this armor menace. This news passed verbally throughout the company NDP. "Arc Light" was code for a United States Air Force B-52 strike. These heavy bombers flew out of Guam and Thailand. They carried out carpet-bombing missions from above 30,000 feet.[227] Word soon came over the radio that the huge bombers were overhead. Charlie Company's NDP was "danger close" to the mission's new bomb drop. "Danger close" indicated that the NDP was less than three kilometers from the mission's target[228] and therefore we risked becoming collateral damage from the strike.

The bombers already were in the air. There was no time for the company to relocate to a safe distance. The grunts sat quietly in the blackness waiting with curious anticipation mixed with a tinge of anxiety. If the Air Force made the slightest error in their calculations, we could be obliterated out of existence. Soon the massive bombers unleashed their payload. No one heard the planes fly overhead, but their bombs struck in a breathtaking exhibition. One minute the night was so dark a person could not distinguish his own hand an arm's length distant. The only sounds were the peculiar noises of jungle wildlife. Then the next minute bright flashes lit up the whole area as if it were midday. Next a brief steady rumble of loud explosions preceded the ground shaking beneath everyone's feet. *Awesome!* Then silence. As suddenly as it began, it was over and darkness again swallowed up the NDP. More important was the silence. No more

---

227. *Encyclopedia of the Vietnam War*, 2000 ed., s.v. "ARC LIGHT (B-52 Raids)," pp. 20-21.

228. Since we were less than three kilometers (*i.e.* 9,842.5 feet) and the B-52 flew at 30,000 feet, we were approximately three times closer to the target than the airplane bombing it.

tanks were heard. Everyone breathed a sigh of relief. The law prohibiting American ground forces from crossing into Laos barred us from checking out the results of the bombing. It didn't really matter. That silence was golden. No damage had been dealt to the company. No MIA. No WIA. No KIA. Everyone was okay.

The whole episode only lasted a couple minutes but produced an undeniable exhilaration. My heart was pounding and I was unable to sleep. I remembered Nguyên, a Kit Carson Scout, telling about his experience on the Hô Chí Minh Trail. Kit Carson Scouts were former Vietnamese Communists who had defected under the Chiêu Hôi program and agreed to serve with American combat units. The Chiêu Hôi (Open Arms) program was a South Vietnamese propaganda effort to encourage defections from the VC and NVA. The South Vietnamese government promised Communist deserters clemency, financial aid, free land, and job training if they changed sides. Nguyên had been an NVA replacement. On his march south from North Vietnam B-52 bombers annihilated his unit. The devastation convinced him that the American military could never be defeated. He immediately searched out an American unit and surrendered. After witnessing a raid that night, I could better imagine the terror he felt when he was in the target area. With this episode I had personally witnessed close air support by a World War II era dive bomber on Purple Heart Hill and B-52D Stratofortress strategic bombers on the Laotian border.

Lam Son 719, the ARVN incursion into Laos, largely was a dismal failure. Their withdrawal ended the Dewey Canyon II mission for the Rakkasans. During the Lam Son 719/Dewey Canyon II operation, 253 Americans had been killed. Another 1,149 had been wounded. At least 5,000 South Vietnamese had been killed or wounded. Another 2,500 were missing.[229] Now American forces

---

229. *Encyclopedia of the Vietnam War*, s.v. "Lam Son 719" by Earl H. Tilford, Jr., p.

began withdrawing from their forward operating camps along QL 9 and pulling back to Quang Tri.

At one point, Charlie Company drew the task of rear guard on QL-9 for the withdrawing Allied forces. That night we were informed that no Allied troops were west of us. Any force coming from that direction would be hostile. I cannot corroborate the veracity of this information, but at the time the grunts in Charlie Company accepted it as fact. In addition, the grunts in the company were told that we were outside the range of the nearest artillery. The company's primary support would be Air Force F-4 Phantom fighters from Da Nang. This was less encouraging. While the men were extremely grateful to have the decisive firepower the warbirds unleashed, their response time would be longer than artillery. The difference was only a matter of minutes, but in combat minutes can be the difference between life and death.

Battalion determined that the rear guard company needed to send out a five man fire team to ambush lead NVA units moving east on QL 9. As an early warning device, sending out a fire team to set up a listening post was a sound military tactic. The difference between an ambush and a listening post was the mission. Ambush implied attacking the enemy; whereas a listening post would withdraw before the enemy could engage the team. *This is insane*, I thought. A fire team seemed entirely too small for an ambush in the present circumstance. The memory of those tanks was still fresh in our minds. *The entire North Vietnamese Army will be rolling east along QL 9. A fire team ambush mission will be suicide. They are using tanks!*

Nevertheless, orders were issued and first platoon was selected. We had no choice. Four grunts were selected. A RTO was needed and I was chosen. We set out before sunset. However, we had no

---

218.

intention of wasting our lives needlessly. So we found a spot off the highway with dense undergrowth. "Wait a minute" vines were extremely abundant. These vines derived their nickname by the difficulty grunts experienced trying to get through them. The vines here were supplemented with "beaucoup" briars. We pulled the vegetation and thorns apart and plunged into the thicket. The last man pulled the plants back into their original arrangement. We cleared enough space for five grunts to lie down. To kill time until sundown we broke out a deck of cards for a friendly game of Spades. Four guys played while one man kept guard.

Around dusk we heard movement to our west. *Oh s***! Here they come. What the h*** are we going to do?* The five of us dropped into the prone firing position, rifles pointed in the direction of the sound. Thumbs quietly flipped selector levers off "safety." Muscles tightened as the noise increased and moved closer. No one dared fire. The sounds suggested a large unit. Firing at it only would precipitate our death. The movement stopped a few meters from our position. We could hear them beating down bushes, but heard no talking. This was not suspicious. The NVA practiced sound noise discipline when in close proximity to American or Allied units. If pursuing the Allied retreat from Khe Sanh, they would realize American units were nearby.

Our plan to use the vegetation as an early warning device had backfired. Instead of providing safety it imprisoned us. We could not withdraw or call in on the radio without our noise tipping off the enemy. Death or capture now seemed certain. *How stupid could we be?* Although we all were experienced combat infantrymen, we had committed a fatal rookie mistake. Fear had led the team to set up in an untenable position with no escape.

Obviously we did not sleep that night. Finally the nighttime jungle noises were loud enough to drown out a radio transmission. I softly whispered into my radio handset that we had movement to our front. Higher up instructed us to await developments until we could clarify the situation better. My thoughts turned melancholic. *If we had mortars or artillery, they could provide cover fire for us to retreat. F-4 Phantom ordinance is too risky without more definitive information. Besides, bombs likely cost too much to waste on only five grunts. Grunts are easy to replace. The Selective Service just drafts five more teenage boys.*

We lay there waiting for an inevitable clash that never came. At dawn the stalemate had to end. We could not remain in the thicket indefinitely. We must return to the company. The tension was high. Sweat poured from our foreheads. We knew that any movement on our part would trigger fire from the unseen foe. However if we did not move, we might be abandoned in the jungle. Dying or capture seemed our only options. Death seemed a better choice than being a POW. As we arose and started out, to our surprise no RPGs burst into us. No AK-47 bullets sprayed around us. The only sound was something crashing through the jungle vegetation. We broke out of our imprisonment in time to catch a glimpse of our nemesis. A herd of Indian muntjac had bedded down for the night near us. Our sudden movement startled them and they bolted for the closest escape. Although it had not been humorous that night, we later laughed at the incident. *Deer, not NVA, had kept us up all night!*

As the withdrawal continued, elephant grass was torched to deny any pursuing NVA concealment vegetation. However on one occasion the fire got out of control and almost burned up first platoon. Fortunately we were able to find shelter in a nearby creek while the flames encircled us. Auspiciously the elephant grass was low at this particular location. The brush fire was not extremely hot, neither was it large enough to cause acute harm. We still gasped for breath while it consumed the oxygen. The smoke irritated our eyes and lungs. However no one was seriously injured.

Finally Dewey Canyon II/Lam Son 719 came to an end on April

9, 1971. General Abrams believed the operation forestalled any major NVA offensive in Military Region I (*a.k.a.* I Corps) during the rest of the year.[230] Charlie Company returned to Camp Evans for stand down. Slicks extracted the company from the bush and flew us to Quang Tri. At Quang Tri, C-47 Chinooks ferried the company to the battalion helipad at Camp Evans. As we disembarked, the 101st Airborne Division band was playing patriotic music. Replacements and REMFs in clean uniforms were standing in formation. The cherries observed apprehensively their new unit returning from the field. The non-combat troops watched with gratitude that they had not been in the bush and also with some trepidation. They recognized the tension the last two months had built up inside the grunts was liable to explode in trouble.

The grunts coming down the ramp of the huge Boeing-Vertol CH-47 Chinook helicopters were in poor physical condition. Over two months in the field had taken its toll on us. Everything was coated in the red dirt of the region. The knees and crotch of my pants had rotted away. So too had the canvas sides of my jungle boots. My socks were caked in dried mud from the night I jumped into the muddy foxhole. They also had dried blood from numerous leeches picked up crossing various streams. Ringworm covered my lower body, from the waist down to my ankles. I also had a boil on the back of my neck. As was customary on stand-down, we ambled down to the Connex that served as an arms room to turn in our weapons and ammunition.

The whole scene was surreal. Filthy grunts unceremoniously walking across the helipad. Each individual's gear and weapon was carried according to his whim. Every step exhibited fatigue from the recent ordeal. Most minds were focused on the clean shower that awaited

---

230. Cosmas, MACV: The Joint Command in the Years of Withdrawal, 1968-1973, p. 336.

their bodies in the olive drab canvas tents set up beyond the Connex. The neat formations in clean green uniforms along the side of the pad appeared nonsensical. The martial music sounded frivolous. The exhausted grunts had rather listen to rock and roll. As they anticipated hot showers and medical exams, thoughts turned to downing a hot meal and getting a night of uninterrupted sleep. It was good to be alive. Contemplations of Pamela and home began to replace the mental detachment that two months of constant combat had generated in me.

Suddenly we spotted a familiar face among the replacements. Cecil had recovered from his wounding near Purple Heart Hill. After being released from the hospital, he returned to the company. He stood in formation wearing new jungle fatigues and grinning mischievously. Several members of first platoon ambled over to him. With a straight face he declared, "D***! You guys stink!" The stoic expression gave way to laughter. Before we could respond, NCOs ushered us away to surrender our weapons. "See you in the mess hall," he shouted as we headed for the Connex and tents.

While on stand-down, the 3rd Battalion of the 187th Infantry (Airmobile) held a memorial ceremony. Nineteen pedestals were lined up in a row. Each pedestal was covered with a poncho liner and a pair of empty boots sat on top. The name of one of the soldiers KIA was mounted on each stand. The entire battalion stood in formation before the pedestals. The observance followed traditional military decorum and pomp. A firing team discharged the customary three rifle volleys as a bugler played *Taps*. The mournful bugle notes and the crack of rifles triggered an unexpected emotion of sorrow — seven of the dead had been in Charlie Company. The KIA included the battalion commander (Lt. Col. Sutton) and his operations officer (Maj. Ron Scharnberg), two men from Headquarters and Headquarters Company, three men from Company D, five from Company E, and the seven from Company C.

Not all was protocol in the ceremony. At one point Bravo Company burst into laughter. Their pet dog defecated on Sutton's poncho liner. Sutton was disliked by some soldiers in the battalion because he drove them hard and would take risks in order to accomplish the mission. The American withdrawal from Vietnam was having unintended repercussions. Without the goal of winning the war, grunts possessed little motivation and enthusiasm for battle. *What was the point in dying now? My death won't affect the outcome. The President has promised to end the war without victory.* At the same time Sutton was willing to get on the ground and share the grunts' dangers. I respected him highly as a good combat leader. I was not alone in my opinion.

During the stand-down, Charlie Company took its customary trip to Eagle Beach. On this occasion the company made the final leg of the trip in a landing craft known as a LCM. The army used these shallow draft boats for harbor service and limited intercostal waterway operations.[231] After watching war movies such as *The Longest Day* in my adolescence, travel in the landing craft recalled the beach landings of World War II.[232] War was no longer as laudable as in the movies however. Fortunately, I was heading to a recreation diversion rather than a hostile beachhead.

The most memorable incident during the stand down was observing a fight with a couple of drunken sailors. They were attending the USO show and made several disparaging comments about the army. By that time the grunts of Charlie Company embraced a cocky attitude forged in the crucible of battle. We believed we were the best

---

231. Stanton, Vietnam Order of Battle, p. 329.

232. *The Longest Day* is an epic war film based on Cornelius Ryan's book *The Longest Day* (Simon & Schuster © 1959), about the D-Day landings at Normandy on June 6, 1944, during World War II. The film was produced by Darryl F. Zanuck and released by Twentieth Century Fox in 1962.

warriors in the world. The sailors were outnumbered and outmatched. I don't remember who swung the first punch. Needless to say, the fight didn't last but a few punches, almost all by the grunts. When the MPs arrived, the sailors were sleeping soundly.

THE FLATS

Two Weeks Leave and Training ARVNs

LEAVE

In early April 1971 it became apparent that the NVA was moving troops, equipment, and other supplies through the A Shau Valley. Operation Lam Son 720 was intended to disrupt enemy logistics in the A Shau Valley. The 1st ARVN Division was sent into the Valley. The 101st Airborne Division provided support for this ARVN incursion into NVA controlled territory. During this period, PF and RF troops were attached to Charlie Company. The PF—Popular Forces—and RF—Regional Forces—were the ARVN equivalent of the American National Guard. The PF were recruited at the village level and operated under the command of an NCO who answered to the village chief and the local ARVN district chief. Normally they operated only in and around their local village. The RF also was recruited locally but was responsible to the province sector chief or the district sub-sector chief.[233]

---

233. 720th Military Police Battalion Reunion Association Vietnam History Project

Charlie Company and the 956 RF Company conducted search and attack operations southwest of Firebase Jack. Jack was approximately 10 kilometers southwest of Camp Evans.[234] The combined force's first contact occurred midmorning, April 26, when it engaged and killed one VC.[235] I missed that firefight. Shortly before it occurred, I returned to the United States for 14 days leave. Throughout most of the Vietnam War, leaves were granted only for humanitarian emergency situations. In early 1971 the military experimented with allowing troops to take a two-week leave rather than an R&R. I was one of the few lucky enough to participate in the experiment. My leave ran from April 21 to May 5. Since travel time was included, I would have approximately ten days at home.

Late in the afternoon of April 20, a supply bird picked me up in the bush and returned me to the battalion area at Camp Evans. I secured my M-16 in the armory and found a cot in one of the hooches. After a restless night—I was anxious to go home—I reported to First Sergeant Herrington. He issued me my orders and travel papers. My field equipment was secured and I headed home to the world, to the land of the big PX. A driver took me to the main helipad in an M151 4x4 utility truck. These vehicles resembled the Jeep of World War II that they had replaced. From there I hitched a ride on a Huey going to Phu Bai. The chopper actually had seats installed. So I fastened my seatbelt and rode like a civilized passenger. It was weird, almost scary to be strapped in after so many CA rides sitting on the floor.

---

website (http://720mpreunion.org/history/project_vietnam/rfpf/rfpf_profile.html). Accessed August 6, 2016.

234. Kelley, *Where We Were in Vietnam*, 5-260.

235. Operational Reports—Lessons Learned, 101st Airborne Division (Airmobile), Period Ending 31 October 1971, p. 1.

I flew military standby from Phu Bai to Tan Son Nhut Air Base in Saigon. This meant rather than board a scheduled flight I would wait in the terminal until space was available on an Air Force plane heading to Tan Son Nhut. While waiting I met two other grunts from the 101st who were also headed home on leave. Fortunately, the wait was brief and by early afternoon we walked up the rear ramp on a C-130 and located available seats around its cargo load. The trip south was uneventful.

We found housing in a transit facility on the military side of the airport. That afternoon we walked over to the Air Force Officers Club for the meal described previously. The next morning we changed out of our jungle fatigues into the khaki tropical army dress uniform and packed our duffel bags. Then we headed over to the main civilian terminal of Tan Son Nhut International Airport. I exchanged $150.60 in MPC for genuine United States dollars. I also had to go through Vietnamese customs. I was concerned that they might confiscate the NVA belt I captured near Purple Heart Hill or the two rounds from the AK-47 chamber and magazine. I had not had time to fill out the paperwork for the belt and the bullets were contraband. I figured if I tried hiding the belt, they surely would find it and seize it. I conceived a bold strategy instead. I simply buckled the belt around my waist as if it were part of my uniform. Surprisingly no one questioned me about it. I removed it when I reached CONUS and stuffed it in my duffel bag.

From Saigon, I flew Pan Am Airlines to Los Angeles, California. The plane stopped in Yokota, Japan. The flight to Yokota lasted about five hours. Then another ten hours was required to reach Los Angeles. Being confined to an airline seat for so long made the trip wearisome and endless. I tried to sleep but my internal anticipation was too much.

In my pocket I carried the last letter I received from Pamela before leaving Camp Evans. I read it dozens of times on the flight across the Pacific.[236] Each time I pulled it from my pocket I scrutinized the envelope. The white Air Mail envelope was framed in a red, white, and blue barber pole design. A red ten cents postage stamp was affixed to the upper right corner. Mostly I contemplated the red ink cursive handwriting. Her script was so familiar. It was the same as every other letter she sent me. They were all special because the inscription was penned in her handwriting. Staring at it connected me to her in some imaginary mystical manner. It was all I had until... *In a few more hours I would look into her beautiful blue eyes! What is taking so long? Can't this jet airplane fly any faster? Surely my watch is running too slow.*

Oh how I pined to see her again. I had asked her to meet me in Hawaii for my R&R, but she refused. Free love may have been in vogue, but Pamela still held strongly to old-fashioned Christian values. She recognized the carnal temptation would be irrepressible if she came to Hawaii and she was unwilling to compromise her morality. At home we still would be tempted, but influences for preserving chastity were stronger there than if we were alone half way around the globe. I would have to wait until marriage for some pleasures.

Flooded with these emotions I pulled out a single sheet of lined stationary. The same familiar script, scrawled again in red ink, coated the page. Slowly I began to read

Dearest LeBron,

I almost could hear her voice calling my name...

Hello love. How are ya? It sure will be great to be with you again. Today was beautiful you would have loved it. I hope it is a pretty when you get home...

I came to the end of the page and flipped it over.

---

236. This letter is the only existent correspondence I have from her while I was in Vietnam. It is dated "10:14 p.m., Friday, April 9."

Darling I love you! I'm about to fall asleep so I had better close. ~~*Thake*~~ *Take care & see ya soon.*

Her spelling and lack of punctuation made telling me that she was sleepy unnecessary. This wasn't one of her long informative communications. It was just a short note penned because she loved me and wrote me virtually every day. At the bottom of the page in a hybrid cursive and print scribble were two words every grunt longed to hear, "Welcome Home!" The exclamation point was an inch tall line with a large circle beneath the line.

Sadly, most people were not so glad to see me. I was dressed in khaki dress uniform. The trousers were tucked in my jungle boots and a 101st Airborne Division patch hung off the pocket button. The heckling of a few civilians in the airport initially caught me off guard. I ignored them and walked briskly to the gate for an Eastern Airlines flight to Atlanta. Pamela and my parents were to pick me up in Atlanta and drive me home. Thoughts of seeing all three of them subjugated any hostility towards the anti-war protesters. Still the jeering hurt. I was not responsible for this unpopular war. I was too young to vote when it started. I was simply doing what my country directed me to do. I was one of 648,500 men drafted into the armed forces and sent to Vietnam.[237] Nevertheless, I was proud to be a member of first platoon, Charlie Company, 3rd Battalion of the 187th Infantry, 101st Airborne Division (Airmobile). Their jeers may have been aimed at the war, but to me they attacked the finest group of men I will ever know, my brothers in Charlie Company. These young men would gladly die for each other. "I would die for you" was no idiom for these guys. On Purple Heart Hill they had proven their loyalty to each other was real and deep.

---

237. Draftees made up only 25% of the troops who served in country. In World War II 66% of American armed forces were draftees.

In those days, people met incoming flights at the gate where the airplane unloaded. I stepped off the plane and walked up the bridge expecting Pamela, Mom, and Dad to greet me as I entered the concourse. They were not there. At first I assumed something must have detained them. So I waited. We didn't have cell phones back then, so I couldn't call or text. Alternating between pacing, sitting, and standing, five or ten minutes passed. I was more than a little disappointed when they still did not show up. Then I decided to check in the terminal. I walked to the lobby, looked around, and saw no one that I knew. Dejected, I moved aimlessly about, finally going outside to the curb, hoping their car was there.

By this time I felt abandoned by those I loved the most. Rejection by antiwar protesters had been disheartening. Now abandonment by my family made me near suicidal. *I thought they would never let me down. Even Pamela had deserted me!* I didn't know what to do. *Go straight back to Vietnam? How do I do that?* I went back into the airport terminal and started up the escalator. I was about half way up when I saw Pamela coming down the escalator parallel to the one on which I going up. To this day I don't know how she did it, but she jumped over the escalator railing and into my arms. Our lips met and we kissed each other all the way up. The movie moment faded when we reached the top. She looked me in the eyes and informed me that I smelled rancid. The odor of Vietnam saturated every cell in my body and I could not wash it away. *Oh well. At least she still loved me.*

My parents soon caught up with us. Eastern Airlines had changed the gate where my flight landed, but no one informed them about the change. They had been as distraught as me. Tears started to flow on both of their cheeks when they laid eyes on me for the first time in eight months.

On Friday evening I drove my Chevy Nova to Pamela's house. *Wow! Driving my car delivered a moment of freedom. I wasn't enslaved to the agenda of the green machine* [slang for the army] *and those Huey helicopters. I could go where I wanted, when I wanted.* I anticipated picking Pamela up for dinner and a movie. Naturally she was still getting ready when I arrived. I sat down on the couch beside her father and waited. He was watching the evening news on television just as he did every night. His world and my world were about to collide in an unforeseen comical moment.

In the era before cable news, Americans' primary source of news was the three major networks' nightly news programs. Every weekday ABC, CBS, and NBC each broadcast a thirty-minute national news program at 6:30 in the evening. I was looking forward to an evening with Pamela and did not pay attention to the television screen.

As I sat on the couch, the broadcast reported on the war in Cambodia. At the time the Cambodian government was seeking to expel all Communist Vietnamese troops from Cambodia.[238] Both sides were armed with the AK-47 assault rifle. The fighting was intense and the news reporter on the screen was at the front. Suddenly I heard the distinctive crack of AK-47 fire from the television broadcast. Kak kak kak kak. Instinctively I dropped to the floor. I was lying between the coffee table and the couch when I realized where I was. I looked up at Mr. Coffman. I was considerably embarrassed. He was staring at me anxiously, undoubtedly wondering if it was safe for his daughter to go out with me. I quickly got on my feet. Before either of us could say a word, Pamela came to my rescue. She quickly ushered me out the door and into my car.

For approximately ten days I soaked up life in "the world." It seemed so illusory. The quickness with which I readjusted to its routines surprised me. But even more surprising was the realization that

---

238. *The Encyclopedia of the Vietnam War*, 2000 ed., s.v. "Cambodia" by Arnold R. Isaacs, pp.54-55.

I didn't belong in these routines. Part of me, the grunt in me, remained in Southeast Asia with first platoon, Charlie Company. No matter how hard I tried, I could not escape the fact that I would soon have to go back.

My return flight to Vietnam required me to report to the Pan Am desk in Los Angeles by 10:00 p.m. on May 2. Since the reporting time was Pacific Standard Time, I was able to attend church that morning. So I dressed in my khaki uniform that morning and picked up Pamela. We drove the half block to Green Acres Baptist Church. We had attended worship there the previous week, but that first service seemed more of a novelty to me. Listening to the choir or to the pastor was more like watching a movie than worship. I had not been in church since September 6, 1970. Unlike the passengers in Los Angeles, the people at Green Acres Baptist Church embraced me…and my uniform. Now, with a keen awareness that in less than 36 hours I would be back in Vietnam, my participation in the worship service took on a fresh meaning. Vietnam, and the danger there, no longer was an alien unknown. This time I knew all about where I was going.

After lunch my family drove me to the Atlanta airport. At the gate I hugged and kissed Daddy and Mother and told them goodbye. Then I embraced Pamela. This time my kiss was no longer short and nervous. I was very much conscious that I would not kiss her again until September. Finally I released her and headed down the ramp to the plane. I glanced back over my shoulder several times before disappearing inside the cabin of the airplane. *Next time I won't go back to Nam! Next time my tour of duty will be over. Man! I can't wait until I get back home for real.* I sat down, gazed longingly out the window, and in my mind I relived every moment I had spent with Pamela during the leave. *I love you Pamela Jean Coffman! When I get back I am going to marry you!*

Training the ARVN PF

While I was on leave, Charlie Company moved into an area known as the Flats. The Flats derived its designation from its relatively level topography. Between Camp Evans and the village of Phong Dien, Vietnamese farmers cultivated tidy rice paddies. Elsewhere elephant grass, bamboo thickets, and intermittent wooded streams dominated. The uniform landscape was broken by occasional hills that rose up like islands in the sea. Previously this area had been called "the flat lands,"[239] but in 1971 the shorter nickname prevailed in the vocabulary of Charlie Company's grunts.

The region was bordered by the South China Sea on the east and a mountain range on the west. Camp Evans was located along Route 601 in the heart of this coastal plain. With only a few exceptions, the civilian population of the region was confined to a strip of land on either side of the road and to the district between QL 1 and the sea. West of Evans, this flat terrain continued until the first ridgeline of the Annamite Cordillera ascended abruptly to create a natural wall. The Annamite Cordillera was a precipitous mountain range that ran north-south virtually the entire length of Vietnam. Triple canopy jungle of teak, mahogany, and other hardwoods covered most of this mountain range. Its steep slopes reached up to peaks 5,000 feet high. It was penetrated by numerous deep gorges with white-water rivers flowing toward the sea.[240] The NVA and VC used the rivers to infiltrate the civilian hamlets in the flats. This new AO meant we probably could expect more hit and run tactics by VC than standup battles with sizable NVA units.

During late April, the entire month of May, and early June,

---

239. Wiknik, *Nam Sense*, p. 11.
240. Kelley, *Where We Were in Vietnam*, p. 5-15.

Charlie Company was assigned the dual task of training RF companies from the Hue region and providing security for Camp Evans. Theoretically, our training procedure involved pairing a rifle squad from the RF company with a squad of Americans. The two squads would function as a single unit, enabling the RF troops to gain combat experience while being trained in American combat techniques. About every three weeks a new RF company would rotate in to be trained. After ten days to two weeks, the RF unit returned home and Charlie Company operated autonomously for a week or so.

The approach was less than satisfactory. Joint operations uncovered a few small caches and killed a small number of VC. Two ARVN soldiers also died. A small number of ARVN and American troops were wounded. The brief timeframe a RF company spent with Charlie Company did not allow for the development of relationships. Without the trust inherent in these relationships, cultural and sociological differences between the Vietnamese and the Americans clashed. American grunts disillusioned with the war resented the perceived apathetic attitude of the ARVN soldiers.

The RF did not always comprehend American military tactical procedures. Nor did they understand Americans. To the average Vietnamese citizen, even those supporting the Saigon government, America looked much like other foreign powers, with little difference from the Chinese, Japanese, and French who previously occupied Vietnam. For GIs, the war was a limited abnormal event in their life. They arrived on a certain day and were scheduled to leave exactly one year later. In stark contrast, these Vietnamese RF soldiers had lived in a war zone their entire life. For them, war was normality. Their apparent indifference may have been shaped more by fatalism than by apathy. Confined together, clashes between the two cultures were unavoidable.

On one occasion, we were startled mid-morning by loud

explosions. Two RF soldiers were fishing by tossing hand grenades into a nearby stream. The concussion left a few fish floating on the surface of the water where they easily could be retrieved. Even more fish sank to the bottom of the stream where they too could be gathered up. We viewed their behavior as a serious breach of security, disclosing our whereabouts to the enemy. They thought of it simply as getting their next meal.

In reality, the enemy likely already knew our position. Hiding an infantry company was virtually impossible in the flats. Nevertheless, the incident called attention to the incompatibility of the two Allied units. Charlie Company was still a good combat outfit. Its soldiers accepted the discipline necessary for success over the NVA, not because they were "gung ho" warriors, but because it gave them the best odds for most of them to survive. On the other hand, many RF soldiers were less enthusiastic for battle than the G.I.s. Like the G.I.s, survival was their paramount concern. But for them, duking it out with NVA regulars was not necessarily their best option. In truth, many did not care if the central government was in Saigon or Hanoi. They simply wanted to grow their rice and live in peace. That would only become possible when the war ended. Hence they just wanted it over. It mattered little to them who won. They didn't think life would change much under either government.

Significant differences also existed between the Vietnamese and the American attitudes concerning property rights and speech. The RF soldiers largely were impoverished farmers. By their standards, the American GIs lived a lavish lifestyle. They saw no problem with taking something they believed the Americans easily could replace. The Americans on the other hand considered this stealing as completely unacceptable. At the same time, the Vietnamese were shocked at the profuse profanity of the Americans.

The cultural clash finally reached its boiling point. Conditions were ripe for a tragic explosion. Two helicopters picked up all the company's officers one morning and flew them back to Camp Evans for a debriefing on the effectiveness of the training. This left a staff sergeant as the ranking soldier at the NDP. The relationship between the RF Company being trained at that time and Charlie Company already had fractured. The two units no longer integrated on a squad level. Instead the ARVNs had half the perimeter and the Americans the other half. The fuse for the morning trouble was a hunting knife. Grunts frequently carried large knives. These knives were not issued weapons, but rather items purchased through commercial sources. They were an expression of individuality and personal pride. One grunt owned an expensive American-made hunting knife that his family mailed to him after he arrived in country. On this particular day, it disappeared. Presuming it stolen by the Vietnamese, he grabbed his M-16 and headed for the PF side of the NDP. His rage was evident. Quickly other grunts grabbed their weapons and joined him. Violence appeared unavoidable.

Fortunately an ARVN translator intercepted the grunts before they commenced their private war. The translator was a good soldier and well respected by the grunts. He requested an hour to locate and return the knife. True to his promise, an hour later he returned with the missing knife. He also carried several other stolen items. Independent of this incident, the experiment to train PF troops was terminated by the Army. When the officers returned from their meeting, they announced we no longer would be training PF troops.

COMPANY RTO

During the days without RF units and after the program was suspended, Charlie Company's role shifted from instructing novices to screening Camp Evans. Our patrolling was intended to disrupt enemy activity near the base and thereby hamper attacks against it. The open, flat terrain precluded hiding our position. Likewise the sparse, short grass made detection of enemy forces equally straightforward. Therefore most enemy movement was confined to the hours of darkness. Charlie Company's security was enhanced by patrolling the area around its NDPs. These small patrols were advantageous, but monotonous. By sweeping several kilometers outside the company's perimeter, they extended our sight significantly. NVA and VC impotence in the area decreased potential contact with the enemy. Consequently life inside the company perimeter was mundane and casual. Poncho hooches were erected in our NDPs to shelter against the hot tropical sun.

One day a loud shrieking sound interrupted the afternoon tranquility. The shriek terminated in a loud bass explosion, sending shrapnel whistling into several unoccupied hooches. Several grunts were standing out in the open near the point of impact. We hit the ground immediately. Initially everyone thought it was a VC mortar attack, but the blast seemed too large for an 82mm mortar round. After an urgent radio inquiry, it was discovered that the round was a 105mm artillery shell fired by an American battery at Evans. Apparently this short-round was only a defective shell. Fortunately, no one was injured in this friendly fire incident.

Sometime during this period, Lieutenant Metcalf was promoted to Captain. While Charlie Company was in the Flats AO, he moved me out of first platoon to serve as a company RTO. When I joined the CP, the two RTOs were both short-timers. Consequently, I had to learn the job quickly. A few weeks afterwards, I was elevated to head RTO. A company RTO retained even less freedom of action than most grunts. He was a slave, chained to the CO by the radio on his back and its six-foot-long handset coiled cord. The antenna ascended up from his back like a large billboard screaming, "Shoot me! Shoot me!" The head RTO literally was the CO's voice. Therefore he always stayed within a few feet of the CO. Every time the CO moved, I also moved. Wherever the CO went, I went too.

As head RTO, I carried the "PRC-35." This radio weighed approximately 35 pounds and was classified as "top secret." Consequently nothing appears in publications on Vietnam equipment. Essentially the radio was a modified PRC-25 that scrambled the radio signal. It worked by inserting a black encryption device into the base of the radio. The device's cypher was changed daily. A transmission on this radio scramble net only could be understand if you were listening to another "PRC-35" on the same radio frequency and had an encryption device set on the same cypher. Many radio transmissions were sent in code using the military phonetic alphabet. This code was changed almost daily. The head RTO was responsible for communicating the revised code to the other RTOs in the company.

Much of what I did was standard army procedure. However, Captain Metcalf also had some unorthodox ideas. One of Metcalf's notions concerned change of command while the company was engaged with the enemy. If he was killed or seriously wounded in action, he wanted me to carry on as if he were still commanding the company. His logic being that every platoon LT, NCO, and RTO knew my voice. They automatically assumed any command that I relayed was issued by him. Thus if I continued to issue orders after he went down, tactical operations would not be disrupted while the company was engaged in combat with enemy forces. Furthermore, it was not uncommon for the NVA to monitor American radio traffic. If they discovered a unit's CO was down, they might exploit any uncertainty created by transfer of command. RTOs routinely made lesser command decisions. Metcalf's mandate involved prolonged and more comprehensive authority. Consequently, Captain Metcalf schooled me in his tactics and combat thinking. Metcalf's philosophy no doubt was derived from the Airborne school of thought that the man on the spot must possess the authority to act.[241] In both Airborne and airmobile combat, causalities can play havoc with the chain of command.

Delegating such responsibility to an enlisted man was based upon Metcalf's comprehension of the men under his command. He knew his soldiers and learned each man's potential and capabilities. He was aware that the army previously had identified me as prospective officer material. He observed me in the field before and after becoming company RTO. Thereafter he tested me with on the job experiences.

On one occasion while Metcalf was on a RIF with second platoon, a RIF from third platoon engaged a NVA patrol. In Metcalf's absence, I took command and issued orders as if he were next to me. I was pretty sure that he was monitoring me on his RIF's radio (and he was). Everything went fine until Lieutenant Colonel Steverson, the battalion commander, arrived on the scene in his C&C chopper. After the engagement was completed, he landed his helicopter to talk with Metcalf, which of course was a problem since Metcalf was elsewhere. I hastily explained that the CO was with another RIF and had been issuing orders through them. Then I added that apparently their radio had just gone out. I'm not sure he believed my cover-up version, but I never heard of any repercussions from the incident and Metcalf did not modify his thinking about the procedure. That evening he held a debriefing with me concerning the firefight. He was pleased with my performance overall.

For me, the incident provided a little narcissistic satisfaction. For a fleeting instance, I commanded an infantry company in battle. I also will admit that it gave me some insolent glee by fulfilling an

---

241. Harold G. Moore and Joseph L. Galloway, *We Were Soldiers Once...And Young* (New York, New York: Ballantine Books, 1992), p. 36.

enlisted soldier's fantasy, the reversal of the chain of command. An enlisted man—namely me, a Spec 4—was giving orders to officers and NCOs.[242]

COBRAS ATTACK DOWN THE OLD FRENCH HIGHWAY

Metcalf employed another unorthodox tactic for interdicting NVA infiltration through the Flats. By 1971, the VC military capability was marginal. The NVA shouldered the bulk of the enemy's war effort in the region around Camp Evans. Consequently, the NVA periodically sent significant troops into the Flats to reinforce the VC. Metcalf recognized that these enemy units easily could avoid the customary platoon or company size American units operating in the area. So he divided Charlie Company into five-man fire teams with instructions for them to set up ambushes. He dispersed these teams in a square grid pattern with the company CP in the center of the configuration. Fire teams were spaced approximately one kilometer apart, taking advantage of whatever cover and concealment that was available. This deployment provided for mutual support by locating every fire team within one kilometer of at least two other fire teams.[243] Metcalf also maintained a small reaction force at the company CP's NDP. Hence in the level terrain of the AO, each fire team could be reinforced quickly. The dispersal saturated an area approximately 16 square kilometers (about ten square miles) or more with American troops, thereby disrupting NVA traffic in that area.

Members of the company CP monitored our radios 24 hours

---

242. Spec 4 = Specialist, Fourth Class, an E-4 rank.
243. 1 kilometer = 0.62 miles.

a day. At night, enlisted men and officers alternated radio watches. Whoever was on watch listened to all of the radios in the CP.[244] During the day, if the company was on the move, each RTO monitored his own radio. Whenever the company remained in a stationary position, a small speaker often was set up to broadcast incoming calls. The volume could be adjusted so that the RTOs could move about in the CP and still hear their radios. If no one was talking over the network, the speaker transmitted a muted crackling noise. When any RTO squeezed the handset on his radio, it interrupted the normal static and alerted all RTOs that someone on the network was about to speak.

Late on the afternoon of May 17, I was relaxing in the CP hooch after a quiet and mundane day. To date, Captain Metcalf's interdiction scheme had failed to generate any contact with the enemy. I was debating with myself about the evening meal. *Do I want the ham slices or the turkey loaf?* Both of the CP's radios were setting on the ground just outside the poncho hooch. One radio's frequency was set on the company network. The other was on the battalion's frequency. The almost imperceptible sound of static assured me that all was well. Abruptly telltale silence on the company frequency telegraphed an unexpected transmission was coming. A fire team from third platoon called in an early sitrep.[245]

"Six, this is Three Five, we have dinks in the open, over."

I shot up and grabbed the radio's handset. Metcalf heard the message and bounded over to where I squatted. He pulled a map out of his trousers' side cargo pocket and located Three Five's position on the

---

244. Normally three radios were in our company CP. One radio was set on the company's network to monitor the platoons in the company. Another radio was on the battalion network and a third radio belonged to the artillery FO.

245. Radio jargon for "situation report."

map. He looked directly at me and said, "I want clarification about the number of enemy troops sighted."

I said softly, "This is Six.[246] Verify number of dinks sighted, over."

The static immediately terminated, a cue that another report was forthcoming. The voice on the other end of the line quietly replied, "Six, this is Three Five. I have three, no four, over."

He paused momentarily. I could feel my heart rate increasing in anticipation of the next report. I sat on my heels next to the olive PRC-25 with its black handset at my left ear. By now the junior RTO had joined us and sat next to my PRC-25. Captain Metcalf stood stoically across the radio from me. Unless the enemy troops spotted our fire team first, Metcalf would determine what occurred next. I looked up at him and waited for the next radio transmission. He seemed perfectly serene. His demanding eyes were looking at me. I could sense that he wanted more information.

When radio transmission was resumed, the calm in the voice on the other end had vanished. It now trembled as the RTO continued, "Six, this is Three Five. Wait one, now's there's at least nine or ten. S***! They're still coming!"

The alarm in his voice was unmistakable. Then he went silent.

I tried to raise the mute RTO again, but to no avail.

"Three Five, this is Six. Sitrep, over."

Nothing. Just the radio static…only now the static sounded loud

---

246. "Six" was the standard radio designation for the commanding officer. Customarily a unit designation preceded the "six." For example, "Charlie six" denoted the company commander of Company C. "One six" designated first platoon's platoon leader. Since this call was on the company network the use of "Charlie" would have been superfluous and so was omitted. Since the RTO spoke for the CO, he used the CO's call sign. If a distinction was needed, the CO was designated "six actual" and the RTO "six romeo."

and disturbing. *What has happened to Three Five?* The radio static suppressed the answer to the question. I softly whispered, "Three Five, break squelch if you cannot talk. I say again. Break squelch if you cannot talk."

Breaking squelch involved pressing the talk button on the radio handset. This created a click, then a hiss on other radios operating on that frequency, but made no sound on the transmitting set. No signal could be heard. Just silence. I glanced up at Metcalf and nodded. I could read the trepidation on his face. What had occurred? We should have heard gunfire if they had been spotted. Could they have been surrounded and captured? The uncertainty of their fate intensified the growing anxiety. The weight of command now was apparent on the Captain's face. His next decision could mean life or death, for not only the silent fire team, but other fire teams as well. Yet he did not know the exact situation. Nevertheless, he alerted the reaction team. They scrambled to put on their combat gear. In a couple of minutes they were ready to move out. Meanwhile, Metcalf began to formulate two plans. One was based on the assumption that Three Five had been captured. The other was based on the hope they were still a functioning combat squad. Although it seemed to last forever, a few minutes later the silence ended.

Finally the voice on the other end spoke faintly, "Six, this is Three Five. We've got at least 30 dinks in the open. They're just standing around. More are coming. Request permission to pull back, over."

"Three Five, this is Six Romeo, wait one."

I looked at Captain Metcalf. He nodded affirmative.

"Three Five, this is Six, roger. I say again, pull your party back to Charlie Papa, over."

"Wilco, out."

Although withdrawal in the face of a superior enemy involved significant risks, the voice on the other end revealed obvious relief.

Despite their best effort, the fire team was detected by the enemy. The two sides exchanged small arms fire and grenades. Two grunts were slightly wounded but made their way back to the company CP safely.

Meanwhile, Metcalf ordered all units to reassemble on the company CP. Within five minutes the fire teams began to appear at the CP. It only took about ten minutes for the team that had reported the enemy troops to reach the CP. Metcalf questioned them directly for specific information. He learned the dinks were NVA and VC. They were carrying AK-47s, some RPGs, and at least one mortar tube. While the captain debriefed the squad, the rest of the company assembled around the CP.

Following the debriefing, Captain Metcalf instructed me to call in a request for aerial artillery support. The CP was located beside an old French road. The enemy force was located on the same road, about two kilometers away. By the time that everyone in the company reached our location, I was on the radio with an inbound team of Huey AH-1G Cobras.

"Griffin One Niner, this is Charlie Six, over."[247]

The voice on the radio pulsated from the vibration of the helicopter. "Charlie Six, this is Griffin One Niner. We are inbound on your paws, over. Confirm your location, over."

"Griffin One Niner, this is Charlie Six. We are at map coordinates lima golf tango papa lima kilo, over."

"Charlie Six, this is Griffin One Niner. I have map coordinates lima golf tango papa lima kilo. Any instructions? Over."

"Roger, Griffin One Niner, this is Charlie Six. Follow the old French highway and come in on the deck. I say again. Follow the old French highway and come in on the deck. Over."

---

247. Griffin was the call sign of Battery C, 4th Bn., 77th Artillery (Aerial Rocket), the only Cobras stationed at Evans.

"Charlie Six, this is Griffin One Niner. Roger. We are inbound now. Pop smoke, over."

Standard procedure for helicopters—both transport and gunship—called for ground troops to release colored smoke to mark the ground force's position. The pilot then identified the color. The process precluded enemy troops from releasing smoke to confuse the situation. I pulled the pin on a violet smoke grenade and tossed it on the road just beyond our perimeter.

I squeezed the handset and said, "Griffin One Niner, this is Charlie Six. Smoke popped. Over"

"Charlie Six, this is Griffin One Niner, I have purple haze, over," the oscillating voice of the Cobra pilot responded.

"Griffin One Niner, this is Charlie Six. Affirmative. I have you in sight. You may fire as soon as you pass our location. Over."

"Roger. I have your paws in sight."

Normally gunships would not pass over friendly troops for three reasons. First it risked hitting the ground troops the mission was intended to support. Second, it allowed the enemy force to bring its maximum fire against the gunships. Third, it denied the helicopter its optimum target, the long axis of the enemy line.[248] However, these precautions did not apply in this case. We were not engaged with the enemy and the gunships would not see the hostile troops until they already passed our position. Furthermore, at last report the enemy target was in a cluster rather than a line of battle. Using the old French highway as a landmark enabled the Cobras to strike the unsuspecting NVA without warning if they still were there.

Some 15 minutes had elapsed since Three Five's brief firefight with the assembling enemy force ended. That was plenty of time for the foe to vanish. On the other hand, they may have dispatched a

---

248. Dunstan, *Vietnam Choppers*, p. 124.

patrol to investigate Three Five's site. If so, it would take the patrol time to organize, recon the site, and report back to the main body. The next few minutes would reveal what option the enemy had taken.

The throbbing roar of the two helicopters' engines announced their arrival. They were flying at 140 knots (161 miles per hour).[249] I barely had time to spot their narrow silhouettes coming toward us. The vegetation mostly was low elephant grass and the two gunships were only about twenty feet in the air as they zoomed directly over my head. Their prop wash kicked up dust and debris. By then they were on top of the enemy force. Rockets flashed from their pods, leaving a white smoke trail. We could hear the rockets detonating on impact, but could not see the actual blast. The Cobra's grenade launchers added smaller detonations. The explosions were accompanied by the low purring of their miniguns. Gradually gray smoke rose above the tree line that obscured our view of the target. The aerial bombardment had ignited the vegetation within the target. I watched in awe as the two gunships circled and made a second pass.

"Charlie Six, this is Griffin One Niner, we just wasted a whole Dink company. We are heading home, over."

"Griffin One Niner. This is Charlie Six. Roger. Out."

By the time the two gunships finished, the sun had set behind the western mountain range. Down the old French road the glow of burning fires was clearly visible. Captain Metcalf took two platoons and moved down the highway in a classic military tactical formation of two columns, one on each side of the road. Flankers spread out beyond the columns. Metcalf was in the center of the road with me tagging three feet behind him. Further down the dirt road we passed the ruins of some masonry French colonial plantation buildings

---

249. Maximum speed in level flight was 140 knots, but in a shallow dive the AH-1G Cobra could go up to 190 knots. Dunstan, *Vietnam Choppers*, p. 111.

silhouetted in front of the flames. The scene was surreal. It was like something out of an old war movie.

The column halted without warning. Metcalf and I sprinted to the point to investigate the reason for stopping. In the burnt grass and on the road were numerous corpses and body parts. The two platoons spread out and swept the area. They did not find any functioning weapons or living enemy soldiers. In the dark it was difficult to determine a body count from the carnage wrought by the two Cobras. We estimated about twenty dead NVA. When we returned at daylight, the NVA already had removed all of the intact bodies save one VC. There was blood, bandages, bits of human remains, and fragments of uniforms and equipment. But, according to 101st Airborne Division policy, an actual body was obligatory for an official body count. I don't know what the Cobra crews reported. The division after action report only mentioned the one VC.[250] Too much emphasis was placed upon body counts in Vietnam at any rate. Nevertheless, whatever plans that particular NVA/VC unit was preparing to carry out that night were definitely thwarted.

* * *

At 6:35 a.m. on the morning of June 7, six 122mm rockets were fired into Evans from an area on the first ridgeline known as Rocket Ridge. Twelve Americans were injured and two buildings damaged. Tube artillery returned fire.[251] From our location in the Flats, Charlie Company was able to provide the exact coordinates of the enemy location to the

---

250. Operational Reports—Lessons Learned, 101st Airborne Division (Airmobile), Period Ending 31 October 1971, p. 3.

251. Operational Reports—Lessons Learned, 101st Airborne Division (Airmobile), Period Ending 31 October 1971, p. 8.

gunners. We were too far away to RIF the area afterwards. Hopefully the artillery responded rapidly enough to inflict hurt on the rockets.

* * *

About this time I acquired a small silk American flag, approximately 12 by 18 inches. The flag was trimmed with gold fringe. As head RTO, I carried a seven-section folding antenna. When the antenna was extended, it was almost ten feet high. I attached the flag to the antenna. When the antenna was set up on the radio, the flag usually floated defiantly over our position. The memory of the flag over Purple Heart Hill had not faded.

On the evening of July 4, we were set up on the first ridgeline adjacent to the Flats. After dark, Independence Day was celebrated with spectacular fireworks. They were not your stateside garden-variety fireworks, but United States Army firepower...81mm mortars, 105mm and 155mm artillery shells, a mad minute of firing. Infantry units in the Flats joined in the celebration. From our hilltop, Charlie Company grunts witnessed an awesome sight as tracers and explosions lit up the darkness. Green tracers occasionally joined the display as panicked NVA responded to the sudden outburst of firing.

The next morning we moved back down into the flats. July 5 was a quiet, sunny day. The CO sent out a couple of RIFs but they did not encounter any signs of the enemy. Most of the company passed the time with horseplay. For the biggest distraction, some grunts staged a medieval joust. Two Vietnamese Kit Carson Scouts climbed up on the backs of two of the largest Americans. Then the two teams raced at each other in an effort to knock their opponent down. Fortunately no one was injured and it provided great entertainment. The whole day felt like we were attending a Boy Scout camp rather than being a combat unit in a war zone.

TYPHOON HARRIET

Heavy rain fell all day on July 6. Camp Evans measured 10.16 inches in a 24-hour period.[252] Severe wind accompanied the rain, making securing field hooches difficult. As the rainwater drained off the mountain range west of our location, water began rising in Charlie Company's NDP. I grabbed six empty wooden ammo crates from the mortar position and laid them side by side inside my hooch. This provided a bedstead 33 inches wide, six feet long, and six inches high. I blew up my gray air mattress and lay on top of the crates to make a cozy bed. As the water level rose higher, I added another layer of crates. When I had to add a third layer, I knew we were in trouble. Nevertheless I wrapped my poncho liner snuggly around me and hoped the water would stop rising.

My M-16 leaned against the foot of my makeshift cot. A sudden gust of wind lifted the poncho over it. The shoestrings securing the poncho corners gripped the cover securely and the poncho started to flap. The flapping poncho whacked the rifle's flash suppressor and the weapon slid along the edge of the crate. Gaining momentum, it toppled into the water. I heard the splash and glanced down where my rifle had been standing. *Oh s***!* I should have gotten the weapon out of the water immediately, but I was wet and didn't want to leave my poncho liner cocoon.

By mid-afternoon the water was 12-14 inches deep and rising rapidly. My makeshift bed began to float and come apart. The current from the rising water was washing away anything that floated. I lost a prized canned ham that Mother had sent to me in a care package. It

---

252. Annual Typhoon Report, 1971: Fleet Weather Central/Joint Typhoon Warning Center; Guam, Mariana Islands, p. 5-43 (http://www.usno.navy.mil/NOOC/nmfc-ph/RSS/jtwc/atcr/1971atcr.pdf).

was becoming increasingly apparent that Charlie Company's position was untenable.

Captain Metcalf resolved to abandon the NDP and head for high ground. The closest hilltop was Hill 61, approximately 10 kilometers away. He ordered everyone to pack their rucksacks immediately. Poncho tents quickly were struck. The wind aided the breakdown but relentlessly hampered packing ponchos and poncho liners. The deep water made recovering small items on the ground almost impossible. Long before the task was completed, everyone and everything was saturated with water. We were extremely miserable.

Everything we could not carry on our backs was left behind. The discarded equipment included 81mm mortar rounds still in their crates. By the time we finished packing, the water was above our knees. You could feel it crawling up your leg. Shortly after we started moving, it had climbed to our waists. A sense of urgency goaded even the most recalcitrant troublemaker to speedy obedience.

That night Typhoon Harriet hit the Vietnamese coast along the DMZ packing winds of 115 miles per hour.[253] Four civilians were killed; 14 others were missing. Approximately 2,500 homes were damaged or destroyed by the storm. Our NDP was at least ten to fifteen feet underwater next morning. Metcalf's prudent decision to move had been correct.

Humping in the dark through waist high water during a typhoon was a treacherous proposition. But the only alternative was drowning. So we kept pushing...all night long. Since we were close enough to Evans for artillery support, a battery fired illumination rounds continuously. The light enabled us to watch each other in case someone

---

253. Spokane, Washington, *Spokesman-Review* July 7, 1971 (http://news.google.com/newspapers?id=De9LAAAAIBAJ&sjid=_uwDAAAAIBAJ&dq=typhoon%20harriet&pg=7223%2C1983428). Accessed August 7, 2016.

stumbled or stepped into a hidden depression. We moved in a tactical column with security on each flank. However, that night the NVA threat was trivial. Like us, the NVA was fighting the weather and did not have time or desire to harass a company of Screaming Eagles. Their underground tunnels and bunkers were no better against floodwaters than our above ground hooches. Almost certainly they filled up with water and became more untenable than our old NDP.

Remembering that my weapon had been submerged for a significant time, I requested permission to discharge the rifle. Captain Metcalf pointed to a small frog swimming nearby and jokingly said that it might be NVA. I took aim, yelled, "Fire in the hole," and squeezed the trigger. The M-16 fired. I missed the frog, but I knew my weapon was operational. Whatever problems plagued the earlier models of the M-16, they no longer impeded the M-16A1. As long as the barrel bore was unobstructed and the firing pin lubricated, the rifle worked.

By daybreak, we had reached Hill 61. We were cold and wet and every muscle in our bodies ached. Our equipment was soaked, adding pounds to the load we carried. The move had resembled a mass migration more than an infantry march. Upon reaching the summit of Hill 61, weary grunts collapsed from exhaustion.

The typhoon moved into North Vietnam and by mid-morning the sun broke through the clouds over Charlie Company. With the improved weather conditions, it was necessary for Charlie Company to recover its combat edge. Therefore the day was spent establishing a new defensive perimeter on the hilltop and drying everything out. Foxholes were dug in the rocky soil. The low vegetation presented natural fields of fire. So clearing the slopes in front of the holes was minimal. Poncho hooches were erected behind the foxholes. Poncho liners were spread out on top of the hooches to dry out. The contents of wet rucksacks were spread out on the ground to take advantage of the sun's parching heat.

The water slowly began draining away. A RIF was sent back to the previous night's NDP to recover the mortar rounds. The American 81mm rounds easily could be fired in an NVA 82mm mortar. Therefore it was urgent that the RIF reach the site before the NVA could scavenge them. Happily, all the rounds were accounted for and removed to the company's new NDP.

A few days later, Metcalf completed his tour in Vietnam. He was replaced as company commander by Captain Thomas A. Rodgers. Despite the grunts' early trepidation, Metcalf had proven to be an excellent combat officer. He was a good leader and had earned the respect of his men. The men who commanded Company C, 3/187th Infantry and its different platoons varied in their ability. Some were bad leaders. They made mistakes and people died. Some were incompetent and never earned the respect and loyalty of their men. One of these expressed it succinctly. "I'm the highest paid private in this army," he said. Others were arrogant and distained by those they led. One of these earned the nickname "Captain Milk and Cookies" because he had fresh milk and cookies delivered to him on almost every resupply. Never did he share or have similar treats sent to his men. Others were good officers and did an excellent job commanding the company or its platoons. Still, in my opinion, Stephen Metcalf was the best company commander of the group. Lieutenant Smith was the best platoon leader that I served under. I am not alone in this appraisal of our officers.[254]

---

254. CSM Herrington had served under Col. Stevenson when the battalion commander was only a company commander. Herrington rated him and Metcalf as two of the best company commanders under whom he served.

DIVISION COLOR GUARD

Shortly after Typhoon Harriet passed, Charlie Company received a special honor. Command Sergeant Major R. J. Dunn from division headquarters had completed his tour in Vietnam. Charlie Company was selected to serve as the 101st Airborne Division Color Company for his departure ceremony. The company was picked up from Hill 61 by slicks and flown back to Camp Evans. Upon reaching the battalion area, brand new uniforms were issued. Correct names and appropriate insignias adorned the uniforms. Mine had my last name, a subdued cloth CIB, a subdued Fourth Infantry patch on the right shoulder, and a colored 101st patch on the left shoulder. A black metal Spec 4 insignia of rank was pinned on each collar. Jungle boots were polished. This was particularly irritating to most grunts. Lengthy service in the bush eroded the black leather finish and left a tan downy texture. This doeskin appearance exemplified an experienced veteran and therefore was a mark of pride to grunts. However grunts were pragmatic creatures. Fighting the NVA or wearing shinny polished boots? No contest! The grunts begrudgingly surrendered the symbolic footwear for tangible time out of the bush.

The next morning, Charlie Company boarded Chinooks and flew to Camp Eagle for the ceremony. The ceremony was held on the runway tarmac. The company formed in two separate formations on either flank of the color guard. The numerous flags of the various units that comprised the 101st Airborne Division made an inspiring sight. The division included ten battalions of airmobile infantry, plus at least 24 other units. Each had its unique banner.

As impressive as the formation was, grunts actually found the small delegation of American Red Cross Donut Dollies far more impressive. Donut Dollies were part of the Red Cross Supplemental Recreation Activities Overseas (SRAO) program. They were all college graduates who volunteered for service in Vietnam. More importantly to the grunts, they were all American girls. The mission of the Donut Dollies was to cheer up American troops in Vietnam.[255] These young women had "round eyes" and wore short baby blue dresses. Most grunts in Charlie Company had not seen an American girl in many months. From a grunt's perspective, these Red Cross field workers were more gorgeous than any Hollywood starlet. When given a few minutes of free time, virtually the entire company crowded around these young ladies. There were too many grunts for more than a cursory look. The Donut Dollies graciously flirted and made small talk with the fortunate few guys closest to them. The rest of the company jostled for better views. Although I was not close on this occasion, later in the day Captain Rogers sent me on an errand. In transit I met one of these girls alone. We talked briefly and I snapped her photograph.

Following the ceremony, the company boarded the Chinooks and returned immediately to Camp Evans. The following morning Charlie Company returned to the bush. The CA was "cold" and life returned to normal, at least the "normal" for a grunt in Vietnam. Major General Thomas Tarpley wrote a letter of commendation thanking the company for "their outstanding appearance and performance" during the ceremony.[256] However to us grunts, the affair's impact lay in moving us another day closer to our DEROS, not in the military spit and polish. On the other hand, looking at Donut Dollies was a lot better than being shot at by the NVA. Hence for the grunts in Charlie Company, participation in CSM Dunn's DEROS was a win-win situation.

---

255. *Encyclopedia of the Vietnam War*, 2000 ed., s.v. "United States Red Cross Recreation Workers" by Lori M. Geist, p. 433.
256. Letter of Commendation (9 July 1971).

FSB MEXICO AO

Final Days in the Field, R&R, and DEROS

HAPPY BIRTHDAY, ZERO!

The company CP included an artillery FO. His job was to coordinate artillery strikes for the company. Our FO's RTO was Geary Mortimer from El Paso, Texas. Geary had been a rodeo cowboy before he was drafted into the army. Whereas most grunts carried photographs of our girlfriend or wife in our wallet, he carried pictures of a bull and a bronco. We non-rodeo riders considered this a bit weird. However Geary said that both animals had thrown him every time he rode them previously. He was determined to get back home and ride them without being thrown. Every grunt needed a future goal in order to cope with the present reality. This objective provided hope. For me it was the aspiration to marry Pamela Jean Coffman. For Geary it was overcoming these two animals. Outside of this quirk, Geary was a really great guy. He and I shared the same birthday.

On our twenty-first birthday, the guys in first platoon showed up at the company CP to help us celebrate. They brought a C–Ration pound cake with a lit cigarette stuck in it as a substitute for a candle.

Geary and I posed for photographs holding the cake while they sang "Happy Birthday!" The merriment was a prime example of the grunts' ability to make the best of their situation by bringing innovative solutions in both combat and their daily life.

RUBBER RAFT PATROLS

One of the more hair-brained schemes I witnessed in Vietnam was using rubber rafts to patrol the small streams in our AO. We were still operating in the Flats adjacent to the first ridgeline. Part of our mission was to interdict local communist sympathizers from aiding enemy troops in the mountains. Lam Son 719 temporarily had disrupted the NVA supply line down the Hô Chí Minh Trail. As a result, NVA and VC units in the area relied on local food sources. The rivers and creeks flowing out of the mountains served as super highways for NVA foraging parties and nearby civilian supporters. The communists from the local population often dropped food at designated supply stations along the banks of the streams rather than make direct contact with the NVA troops. Charlie Company sent RIFs along the watercourses searching for these supply stations.

Someone conceived the ill-advised idea that waterborne patrols would be more efficient in finding these supply stations than hacking through the jungle. So the company obtained a number of rubber rafts. A couple of the vessels were two-man rafts, but the majority of them were one-man types. The vessels were bright yellow, hardly invisible in a green jungle environment. The whole venture was extremely dicey. Patrols using the rafts were dependent on stealth and surprise for survival. If an enemy ambush caught them on the water, they would be sitting ducks. The high risk factor led to only using volunteers for these patrols. I considered the proposition suicidal and never went on one of the patrols. Fortunately, the project was abandoned before anyone was hurt. The rubber rafts proved too flimsy for extended field service and the supply stations were too well camouflaged to be spotted from the water.

A more practical approach to interdicting the enemy supply chain involved establishing checkpoints along these rivers and streams. On one occasion a checkpoint stopped three young females coming out of the mountains. The women were prostitutes who had been providing their services at an NVA base camp. The division sent a helicopter to take them back to Camp Eagle for interrogation.

On an even lighter note, during this period Charlie Company coordinated several air strikes through an Air Force Forward Air Controller. The FO flew a Cessna O-2A Skymaster. The O-2A had two tandem prop engines, one in front and one in the rear of the fuselage. The engine behind the cabin required a twin boom tail design, thereby giving the airplane a unique appearance.[257] It had the ability to fly at relative slow speeds and low altitudes. After one mission, the pilot—who I believe went to college with Captain Rogers—jokingly chatted with me after a successful mission.

"Charlie Six Romeo, this is Bilk Two Three, over."

"Bilk Two Three, this is Charlie Six Romeo, over."

"Charlie Six Romeo, great job. I am heading home. Going to get a shower and hit the club. Let me buy you a drink."

In jest I remarked, "I'll take a Segram's® Seven."

A few days later the FO contacted me, "Charlie Six Romeo, this is Bilk Two Three, over."

---

257. See Cessna O-2A Skymaster fact sheets from the United States Air Force websites for the National Museum of the US Air Force (http://www.nationalmuseum.af.mil/factsheets/factsheet.asp?id=304) and Hulbert Field (http://www2.hurlburt.af.mil/library/factsheets/factsheet.asp?id=3432).

"Bilk Two Three, this is Charlie Six Romeo, over."

"Charlie Six Romeo, I'm inbound to your Alpha Oscar. What is your current paws? I say again, I'm inbound to your Alpha Oscar. What is your current paws? Over."

"Bilk Two Three, I read you are inbound to my Alpha Oscar and need my current location, over."

"Charlie Six Romeo, this is Bilk Two Three. Roger. Over."

"Bilk Two Three, Wait one, over."

I grabbed my map and cardboard decoder. Quickly, I identified our location on the map and transferred the map coordinates to the current phonetic alphabet code.

"Bilk Two Three, this is Charlie Six Romeo. Our current November Delta Papa is hotel oscar mike xray hotel kilo. Over."

By this time I could see a dark spec on the horizon moving almost directly at my position. We were in a wood line adjacent to a large open field. The grass in the field was less than two feet tall. I pulled the pin on a green smoke grenade and tossed it into the field. Green smoke billowed upwards.

"Bilk Two Three, I have popped smoke, over."

"Charlie Six Romeo, I have green smoke, over."

"Bilk Two Three, affirmative. Over."

"Charlie Six Romeo, this is Bilk Two Three. I have a special delivery for you. Over."

I was bewildered. *A special deliver for me? Surely he meant for the CO.*

"Bilk Two Three, this is Charlie Six Romeo. Do you mean Charlie Six Actual? Over."

"Negative! I say again, negative. This one is for you, Romeo."

I heard the pilot chuckle as he circled our position. He made several passes over the field by our campsite. Eventually his plane banked as the plane flew directly over us at a very low altitude. The FO stuck his hand out the side window and dropped something. A small white parachute opened and it floated to the ground. It hit about 30 meters outside our perimeter.

He made one more low level pass and waved as he headed off into the wild blue yonder. Since he did not identify the package's contents, I was oblivious to what it might contain. Nevertheless I thanked him. Then I grabbed my M-16 and headed out to pick up the dropped shipment. Fortuitously, the airdrop had landed in front of first platoon. By the time I made it halfway to the spot, Larry Klag had retrieved the package. He handed it to me and asked, "What's inside?"

"I'm not sure," I replied.

We walked back inside the cover of the woods. The curious airdrop generated lots of interest. An audience quickly gathered to witness the disclosure of the mysterious package's contents.

"Break it up! One peanut butter claymore will wipe us all out," I howled facetiously.

No one moved. I tore open the cardboard box. Packed securely inside the carton was a glass bottle containing a fifth of Segram's® Seven whiskey. Since drinking in the bush was taboo, I carried that bottle in my rucksack for eleven days. When we got back to Evans, I shared it with the members of the CP and some of my buddies in first platoon.

JOYRIDING IN A LOACH

In late July, Charlie Company received orders to relocate into a new AO. The company had operated in this AO recently. Whenever the company moved into a new AO, division headquarters sent the company commander a helicopter for an aerial reconnaissance of the new terrain. However only a few weeks had elapsed since Charlie

Company had been in this same AO and the CO was already familiar with the terrain and other pertinent data. Nevertheless, the customary chopper was dispatched for the CO's use. He thought it totally unnecessary. Army red tape made it simpler to log the flight time than to send the helicopter back unused. Captain Rogers asked if I would like to fly. The helicopter was an OH-6A Cayuse "Loach." I had never ridden in a Loach, so I happily replied, "Yes sir."

The OH-6A was a light observation helicopter. Its front windshield virtually encompassed the entire front of its tear-shaped body. The Loach frequently was used in conjunction with Cobra gunships. During these missions the Loach would fly just above the treetops, using its prop backwash to push aside the leaves so the crew could see beneath the jungle canopy. As soon as the Loach spotted a target or drew enemy fire, the crew marked the enemy location with a red smoke grenade. Then the Cobras would scream in and destroy the objective identified by the smoke. This led me to conclude that OH-6A pilots were "dinky-dau" (slang for "crazy"). My first and only trip in an OH-6A only confirmed my previous opinion of Loach pilots. Actually, my opinion did change slightly after the ride. I now realized one qualification for these pilots obviously was that they must lack a human brain altogether. Loach pilots were not "dinky-dau." They were "beaucoup dinky-dau" ("very crazy!"). The episode also showed me why the CO was so reluctant to go.

The OH-6A radioed it was inbound. I popped smoke. The pilot identified its color and landed in a nearby open field. The rotary wash whipped up dust and debris. I bent over and ran to the chopper, carrying only my weapon and ammo. I climbed into the seat beside the pilot. When I explained the situation, he grinned mischievously. The OH-6A was fast and nimble. Its large windshield presented a fantastic view of everything below and in front of me. There were no doors. So I possessed an equally comprehensive view to the side. Overall the front seat feels more like an open amusement park ride than cockpit of a combat aircraft.

We quickly made a vertical takeoff. As soon the helicopter cleared the trees around the LZ, the pilot squeezed the joystick and we shot forward at maximum speed. After surfing the jungle canopy, we started gaining altitude. The pilot put his bird through a series of maneuvers before diving back to earth. When he began pushing the vegetation back with his rotary backwash, I confess, I grew more than a little alarmed. I kept thinking: *This is dangerous enough when you have Cobra backup, but we don't have any Cobras!* Finally he grew tired of the sport and moved on. Next, he spotted some water buffalos in a nearby rice paddy. Flying about a foot above the ground, the pilot charged directly at one solitary water buffalo in an insane game of chicken. These bovine beasts can weigh over a ton! I'm not certain who blinked first, my pilot or the water buffalo... in all probability the buffalo since I now think it the more intelligent critter. In all seriousness, these Loach pilots had nerves of steel and their courage was invaluable to the war effort. Finally we returned to the company. I hastily debarked from the chopper, glad to be back on the earth where the only dangers were NVA grunts, poisonous snakes, and physical exhaustion!

FSB GLADIATOR

Fire Support Base Gladiator was located approximately 32 kilometers west of Hue. It was one of the numerous fire support bases reopened in April 1971 as part of Lam Son 720. Lam Son 720 was intended to provide a defensive screen to a potential NVA counter-attack during the ARVN incursion into Laos.[258] After months in the Flats, Charlie

---

258. Kelley, *Where We Were in Vietnam*, pp. 5-199, 5-287.

Company was assigned the mission of closing FSB Gladiator. The firebase was located on a mountain peak that provided a spectacular scenic view. Our role was providing security while the artillerymen and engineers dismantled the site. This entailed bunker guard and an occasional RIF into the surrounding valleys. Overall the brief deployment there was peaceful and pleasant.

THE MEXICO AO

Fire Support Base Mexico was an inactive firebase 16 kilometers southwest of Camp Evans.[259] After FSB Gladiator was closed, Charlie Company was inserted into the AO around FSB Mexico. The war had moved away from this old firebase and neither side had posted troops in the area for a significant time. Consequently nature had replenished its flora and fauna. Wildlife was abundant and encounters with various native creatures were common. Sometimes these animals died. Venomous snakes were killed deliberately. Other animals died inadvertently.

Mechanical ambushes posed an aberrant risk to nocturnal animals. One night we were awakened by the explosions from a mechanical ambush. The whole company went on the alert. However, there was no NVA attack. Everything remained quiet; no movement or sound. The next morning a patrol checked out the sight and retrieved a mysterious dead animal. To this day I have been unable to identify its species. The animal was about the size of a small dog. It had a long tail with black and gray rings. Its face was similar to a cat's, but it had a dark mask like a raccoon. The body was more canine, with longer

---

259. Kelley, *Where We Were in Vietnam*, p. 5-332.

legs than a cat. It reminded me of a lemur, but lemurs are endemic to Madagascar. It likely was some species of monkey.

Snakes suffered the most casualties. Vietnam is infested with numerous poisonous serpents, including cobras and various vipers. Grunts usually killed poisonous snakes for obvious reasons. The snake that enjoyed the most feared reputation was the deadly Common Bamboo Viper (*Trimeresurus stejnegeri*). They were bright green and could grow up to about 30 inches in length. Grunts habitually called it "the two step viper" because popular lore claimed its victims died two steps after being bitten. This was gross exaggeration, but the skin around the puncture does turn black and die within minutes after the venom is injected. The wound is very painful and can be fatal if not treated promptly.

One afternoon after humping hard, we paused for a brief rest. The temperature was over 100° F and extremely humid. I started to collapse to the ground propelled by my physical fatigue and the heavy burden of my rucksack—as company head RTO mine easily weighed well over 100 pounds. Whoever was behind me, caught my body in midair with one hand and simultaneously swung a machete with the other. When I looked down, a decapitated Bamboo Viper's body writhed on the ground where I was going to sit. The Good Samaritan flipped the head and body into the vegetation with the tip of his machete.

Some meetings with wildlife were humorous. We were "moving down a blue line"—infantry jargon for following a river or stream.[260] The creek bank was irregular and rocky. Various sized boulders littered the ground alongside the white water creek. The point man was a relative novice and was hopping from rock to rock. Soon he

---

260. The terminology was derived from the use of blue colored lines on maps to designate watercourses.

advanced too far ahead of the column and disappeared from sight. Before we could catch up with him, he came running back screaming, "Alligator! Alligator!"

Alligators are indigenous to North America, not Vietnam. The imagined gator turned out to be a five foot long Water Monitor. It looked at us as we approached, its long gray forked tongue moving rapidly in and out of its mouth. Then the monitor turned and waddled apathetically into the jungle.

Vietnam also was habitat for a large variety of lizards. To the veterans of the war, one lizard indisputably became emblematic of their time in country. It was known as the "F*** you lizard" after the human-like sound it made. Its vocal call mimicked a human voice saying, "F*** you!" The vulgar creature was a Tokay Gecko, the second largest of the Gecko species. It grows to about 11-20 inches in length but vociferously discharges its profane vocal clatter.

The "re-up bird" was another animal whose sound seemed to imitate people. Its call sounded like someone softly saying, "Re-up, re-up." In G.I. jargon, "re-up" was colloquial speech for "re-enlist" in the army. The bird actually was a Blue-eared Barbet, but it always will be a "re-up bird" to those who heard it chatter.

Grunts found the duet of a "re-up" bird and a "f*** you" lizard particularly amusing. The bird would sing out, "Re-up, re-up." The lizard would echo, "F*** you! F*** you!" Troops in the bush, especially draftees, could not help laughing out loud in appreciation of the duet's performance. By 1971, many grunts were not professional soldiers. They were men who answered their country's call in the draft. They merely wanted to complete their tour in Vietnam, go home, and get on with their civilian lives. Few wished to make the army into a career. So the lizard spoke for them when the bird invited them to reenlist in the army.

The couple of weeks we operated around Mexico were more akin to a safari than a war. There was one comical incident that was unrelated to the wildlife. The grunts in Charlie Company had become very innovative in their mechanical ambushes and early warning devices. However, one guy actually became too innovative. He came up with the idea of incorporating a CS tear gas grenade with a trip wire. One day as we were humping through some short elephant grass, the pin came loose and the grenade fell to the ground. Even before it released a mist of white tear gas, grunts were running as fast as they could in every direction. In basic training all of them had experienced the choking sensation produced by CS tear gas. None had been issued gas masks in Vietnam. Fortunately, no NVA were around because the gas fumes chaotically dispersed the company, thereby destroying all military effectiveness. After that incident, the use of tear gas grenades with trip wires was strictly prohibited.

The use of tear gas had been occasioned by an earlier incident that was so incredible if I had not witnessed it with my own eyes, I could not have believed it possible. Charlie Company had walked into a small NVA bunker complex. NVA bunkers were so well camouflaged that they were almost impossible to detect. On this occasion, our point man spotted the bunkers when the column was about a dozen meters away. Some of the bunkers were occupied and a firefight ensued. The company dropped rucksacks and moved out to clear the NVA from their position. The grunts in Charlie Company didn't charge the bunkers as depicted in the movies. Instead they crawled around on the ground, presenting to the enemy as small a target as possible. They endeavored to get into a position where they safely could toss a grenade into the bunker.

One grunt worked his way to one side of a bunker aperture. He was in a perfect spot to lob a grenade into the bunker while lying prostrate. He pulled the pin and tossed the grenade into the opening. Then he hugged the earth waiting for the explosion inside the dugout.

Instead his grenade sailed out of the opening and exploded behind him. Shrapnel kicked up dirt all around his feet and legs. Instantly irrepressible anger seized control over his reasoning. He stood up erect. Then flinging his helmet down, he uttered a few expletives. Next, as if the firefight was a stateside training exercise, he walked steadily back to where he dropped his rucksack. He took out a Claymore mine and marched back to the bunker. Oblivious to the situation around him, he sat down next to the bunker opening. Sitting with his legs crossed, he assembled the mine and tossed it into the bunker. Almost simultaneously he pressed the clacker, killing everyone inside. As soon as his Claymore detonated, the agitated grunt came back to reality. He dropped back to the ground swiftly. Miraculously, despite the heavy firing taking place around him, he was not scratched. The whole episode was so implausible and surreal that had I not watched it unfold I would deny it ever occurred. Yet history is filled with incongruous incidents in combat.

Thereafter, Charlie Company grunts routinely tossed a gas grenade into a bunker before hurling in a frag.[261] NVA bunkers were equipped with a grenade sump, a cavity in the floor similar to a golf course hole. A frag would roll into the sump and the ground absorbed most of the shrapnel. However the sump made little impact on the emission of CS gas. The gas still spread into the bunker, temporarily incapacitating the occupants. Before they could recover, the frag exploded, eliminating the bunker.

During the time Charlie Company operated in the Flats, on a lark, I requested a catalog from the Georgia Institute of Technology.[262] When the catalog arrived in the mail I sent in my application. I knew I was not qualified. I was in the Army due to poor academic

---

261. "Frag" was vernacular for fragmentation grenade.
262. The Georgia Institute of Technology is the official name for Georgia Tech.

performance at Kennesaw Junior College. Georgia Tech required some of the highest academic credentials of any public university for admission While we were operating in the Mexico AO I received my reply. In common English it expressed appreciation for my interest in Georgia Tech but informed me in unmistakable words that I didn't qualify. *In other words I was too stupid to go to Tech. Not surprising…*

The evening before Charlie Company was scheduled to return to Evans for stand-down, I strolled over from the company CP to first platoon's position. Some of my buddies in the platoon had invited me to play cards. Most grunts carried a pack of cards in their ruck and the universal game in Charlie Company was Spades. Gambling rarely was involved; the game's primary function was simply passing away time. Nevertheless, grunts were passionate about the game and competition could be fierce. Periodically, tournaments were held to establish the best teams within a platoon or company. Although I had moved into the company CP, friendships with the older veterans in first platoon remained strong. I frequently spent leisure time with them.

A grunt kept his weapon with him at all times while in the bush. So I grabbed my M-16 and headed over the crest of the hill to where first platoon was set up. I had walked point so much that certain traits of the task remained with me. Foremost among these was the habit of walking with my eyes trained on the ground a few feet in front of me—even today I still stare at the ground while walking. As I started downhill, a low hanging branch caught my face just below my right eyebrow. The cut bled profusely.

Doc said I needed stiches and asked if I wanted to be medevac'd. Knowing that the company was returning to the rear the next day for stand down, I declined. Those troops who came into the rear early typically were assigned to work details. They especially were given priority in assigning the worst details, such as burning human feces.

If I was medevac'd I would find myself on these unwanted details over the next several days. And walking into a tree was not exactly falling upon a grenade. Doc wrapped me in bandages. The location of the cut made stopping the bleeding difficult. When he finished, I appeared to have half my head blown away. I poised for a photograph kneeling with a machete in my hand. I always sent my film home for my family to develop. When time came to send this roll home, I forgot about the picture. When my parents and Pamela saw my bandaged head they freaked out.

With one eye covered my depth perception was lost. Thereafter movements demanded deliberate and vigilant strokes. Occasionally I had to be assisted. Fortunately, the birds arrived early the next morning to fly us back to Camp Evans. I never did see a physician and get stitches. But Doc's improvised head tourniquet was effective. The cut healed rapidly. A few days later I only needed a small bandage and life returned to normal — at least what was normal in Vietnam.

R&R IN SYDNEY

Every soldier was entitled to one week-long R&R while in Vietnam. During in-country processing I had indicated that I wished to go to Sydney, Australia, for my R&R. However since I came home on a 14-day leave I was no longer eligible for the usual R&R. But army policies and military regulations were no obstacle for a company clerk named Dove.

Dove was one of the grunts wounded by a booby trap on Christmas day. He was medevac'd to a hospital in Japan for shipment home. But a clerical error somehow altered his orders. Instead of going home, he was returned to duty with the company. The CO considered his treatment an injustice. Therefore he refused to send Dove back into the field. As an alternative, the CO made Dove the company clerk. The affair left Dove with a "FTA" attitude. "F*** the Army" was a common expression among the troops serving in Vietnam in 1971. Its acronym, FTA, became a passive insurgence against military regulation and a sign the American military was unraveling from the trauma of Vietnam. The letters often were inscribed on helmet covers and flak jackets. It even covertly appeared in the artwork of the Winter/Spring issue of *Rendezvous with Destiny* magazine, an official publication of the 101st Airborne Division (Airmobile). As clerk, Dove thwarted the military red tape and did whatever he could to solicit revenge for the injustice done him. His covert war never minimized his work. He knew his duty and carried it out. But if he possibly could make life easier on the grunts of Charlie Company, he would bend any and every rule. In his chicanery to this end, he was more proficient than Radar O'Reilly.[263] On this occasion I was destined to become beneficiary of his flimflam skills.

While on stand-down, Dove told me that he had read in my file where I listed Sydney as my preference for R&R. He informed me that a slot to Sydney for August 8-15 was available. He wanted to know if I was interested in filling that slot. I replied "yes" but there were several major obstacles. First, I had taken a 14-day leave in April. Soldiers who had taken leave were ineligible for R&R. Second, my DEROS was September 13 and soldiers were not permitted to go on

---

263. Corporal "Radar" O'Reilly (played by Gary Burghoff) was a character of the television sitcom M*A*S*H. The television series (developed by Larry Gelbart) was adapted from the 1970 feature film *M*A*S*H* which was based on the 1968 novel *MASH: A Novel About Three Army Doctors* by Richard Hooker. The series was produced in association with 20th Century Fox Television for CBS. It originally aired between September 17, 1972, and February 28, 1983.

R&R if they had less than 90 days remaining on their tour. Finally, I did not have the money to go.

Dove smiled and said, "I will deal with the first two problems if you can get hold of the cash that you need."

"Okay, I will get back with you later today if I can get the money," I replied.

The prospect of visiting Australia remained alluring. Except for Vietnam, I had never visited a foreign country. This might be a once in a lifetime opportunity to see the "land down under." Still if I was honest, the real attraction was that Australia was not Vietnam. I had less than six weeks left on my tour and so I had developed a serious case of short-timers malady. In the bush I was getting extremely jumpy, too nervous for my own good. An R&R would keep me out of the bush for over a week. With less than a month left when I returned, at most I would have two to three weeks in the bush.

Oh yes! I'm going to get that money and go to Australia.

Acquiring the cash was a two-fold undertaking. The trip would cost about $350.00. I had the money in my bank account in the United States. So first I contacted my father and asked him to wire me $350.00. When I explained why I needed the money, he was ecstatic over keeping his only son out of danger. However, back home in America the bank already had closed for the weekend. It would be another two days before he could send the money. Therefore it would not arrive in time for me to travel to Da Nang and catch the plane for Australia.

Consequently, the second requirement would be to scrounge $350.00 from somewhere in Vietnam. I had less than 24 hours to acquire it and still catch that plane in Da Nang. My fellow grunts in Charlie Company were the only possible source for accruing the funds. I wasn't sure I could borrow that sum of money from them. Like me, most did not keep a large bankroll on them. Furthermore, when I returned I would only have slightly over three weeks left to my DEROS. There was a real possibility that they might never see me again. Would they lend me money under these circumstances?

I asked a couple of friends and they agreed to loan what they could. Before I could ask others, word of my situation spread and others came to me with cash. Sending me on R&R had acquired a life of its own. For the grunts in Charlie Company, raising the money was about something greater than me as an individual. Sending me to Australia had become their special means actually to do something that would make a difference in the war. By August 1971, everyone realized the United States no longer intended to win the war. Hence dying in the war was the ultimate waste. If they could keep one short-timer out of the bush, he had a better chance to get home safely. Consequently "me and my R&R" took on a surrogate role for those entrapped in the closing days of meaningless warfare. Likewise, the violation of so many rules made my R&R a big communal "FTA!" Soon I held $350.00 of MPC in my hands.

When I returned from Australia, the money order was in the orderly room. I cashed it and Dove made sure all of my lenders got their money back. In a letter acknowledging the receipt of his repayment, Geary wrote:

Wow, thanks for the money and I'm glad to help out anytime I can, also I'm glad you had such a good time.[264]

The willingness to loan money under these circumstances says a great deal about the bond and camaraderie that combat forged among Vietnam grunts in Charlie Company, 3/187th Infantry. You may rest assured that the same amity and solidarity was nearly universal among the grunts with other units serving in Vietnam.

Dove managed somehow to cut me travel orders and I headed for

---

264. Letter dated "Sept 10-71."

Sydney. As previously noted, I had two reasons for taking the trip. First, I truly wanted to see Australia and this was an ideal opportunity to do so. On this excursion, Uncle Sam would pay for my travel expenses. Later, all travel costs would become my personal responsibility.

Second, I wanted to stay out of the field. Only one month was left until my DEROS. I was no malingerer. I had put my time in the bush. Other than stand downs, leave, and transfer travel time, I had spent virtually every day in the boonies. I reasoned that I spent more time out there than most soldiers. After all, only one in ten soldiers in the United States Army were infantrymen. Soldiers with another MOS faced their own unique dangers. For example during Lam Son 719 helicopter crews in reality incurred far more casualties than the grunts. Artillerymen sometimes experienced similar conditions to the infantry. But I was short! *Short!* I wanted to live to get home and marry Pamela. Survival and self-preservation now guided my actions. In the bush I was getting jittery. I no longer had that, "It don't mean nothing" attitude wherein dying was a tolerable fate. Instead, I was beginning to believe I would make it out of Vietnam alive.

Dressed in my khaki dress uniform, I signed out at the company orderly room and set off for Australia. I managed to catch a chopper from Evans to Camp Eagle. It already was sundown when the helicopter landed. I was unable to find quarters on this unfamiliar base camp. None of the REMFs I encountered offered any assistance. Their apathy reinforced my dislike for the species. Finally, I did locate a macabre shelter near some army buildings. That night I slept in a gravesite. The Vietnamese circular burial place was about seven or eight feet in diameter. It had a low stone wall around the grave that was located in its center. I figured the wall offered protection from small arms fire and most artillery or mortar rounds unless they made a direct hit on my position. So I curled up inside the wall and listened to someone in the distance playing a tape or record of Three Dog Night singing "Joy to the World."

Jeremiah was a bull frog
Was a good friend of mine
I never understood a single word he said
But I helped him drink his wine
And he always had some mighty fine wine

Singin'
Joy to the world
All the boys and girls, now
Joy to the fishes in the deep blue sea
Joy to you and me.[265]

I gradually drifted off to sleep and did not awaken until the sun's rays began to warm my body. I returned to the airstrip and caught a military standby flight to Da Nang.

In Da Nang I hooked up with a guy named Felix who also was going to Sydney on R&R. We purchased additional civilian clothing at the PX.[266] After clearing customs and exchanging MPC for Australian currency, we waited at the gate for the chartered flight to Sydney. Portable stairs rolled up to the aircraft door enabling us to climb into the airplane. After an uneventful flight, we landed in Sydney.

---

265. Lyrics by Hoyt Wayne Axton, Watts, Christian de Walden, Douglas Gamley, Handel, Isaac Watts, Steve Gatlin, Brian J. Culbertson, Mason. Copyright © 1970 Decca Music Group Ltd., Bleamus Music, K - M Music Inc., Culbertson Music, Universal Music Corp., No Limitations Music, Lady Jane Music. Copied from elyrics website (www.elyrics.net).

266. I already had some civilian clothing from my in-country R&R at China Beach.

In Sydney, a bus transported Felix and me along Darlinghurst Road from the airport to the Kings Cross area. We both rented rooms in the Top of the Cross Hotel. Kings Cross had a sleazy reputation in 1971. The influx of so much American military had led to the growth of numerous businesses catering to their dissolute appetites. The bars Whisky a GoGo and Texas Tavern were among the most popular hangouts for GIs. But the area also had some American fast food restaurants. To a grunt accustomed to the cycle of day and night in the bush, the 24 hour a day hustle and bustle of King's Cross was alien and bizarre.

After changing out of our khaki uniforms and into our civilian clothing, Felix, myself, and two other soldiers headed out to investigate this strange new world. The sun was slipping behind the western horizon, so we headed for the Whisky a GoGo on Williams Street. The popular nightspot already was crowded. Couples bopped on its diminutive dance floor, but the cages hanging from the ceiling caught our attention. Inside each cage beautiful bikini clad girls danced to the rhythm of Motown.

August is early spring in Australia and the locals were anticipating the beaches opening in a few weeks. We were bronzed from the tropical sun and so stood out among them and their pale white skin. This stark contrast had certain perks. Unlike Americans, the Australians were not heavily involved in the antiwar movement. They generally supported the war effort and treated us as heroes. One day I was walking down the street in King's Cross and a pretty young girl came up to me and kissed me passionately smack on the lips. I stood there stunned while she disappeared into the crowd.

We visited the University of New South Wales, the Australian Museum, and the Taronga Zoo and Aquarium. The famous Opera House was still under construction, but we walked to it anyway. At the Museum, Felix and I met two local school teachers. These two young ladies graciously turned into our unofficial private tour guides. They provided lots of interesting information.

At the zoo, the duckbill platypus was on display. This unique creature could only be viewed at certain hours. At the appointed time we went inside the building where the bizarre animal was kept. In the center of the room was a large aquarium. Its water level was about three quarters full of clear water. The aquarium was empty except for the rocks on the bottom and a few logs. We crowded into the room with the other spectators and waited in anticipation of seeing one of the world's most unusual animal species. After waiting for a long time, a door at one end of the aquarium opened. A small brown something darted to the bottom and hid under a log. That was all we saw of the elusive duckbill platypus.

The United States Military had an arrangement with the Sydney City Club that allowed soldiers on R&R to play golf on their course. Early one morning, Felix and I took transportation to the club, planning to play a round of golf. The place was crowded and we were told it would be a long wait unless we were willing to team up with a pair of local citizens to form a foursome. We gladly agreed and soon were paired with a gentleman in his mid-fifties and his 19-year-old daughter. She was very attractive. He was a veteran of the North African campaign in World War II, one of the famous "Desert Rats." After 18 holes, he bought us lunch, a sausage sandwich and lime Kool-Aid®. The week passed faster than I wanted and I soon was back in Vietnam.

There is an old proverb that says, "There is a right way and a wrong way, and there is the army way." Well, the army decided that I was too short to return to the field. I must be kept safe in the rear. The irony of the army's logic concerned the geographical location of the rear. While I was in Australia, Charlie Company was sent to Cam Rahn Bay to act as base security. Cam Rahn Bay arguably was the safest spot in Vietnam. Even the President of the United States visited the base there. On the other hand, the rear was Camp Evans. At that time, Evans had become one of the most exposed bases in Vietnam. So the field was the bunker line at the safest spot in the country and the rear was the bunker line at the most dangerous spot in the country. I would be pulling guard duty in either place, but the army in its wisdom sought to keep me safe by putting me in the most dangerous bunker in Vietnam rather than in the safest bunker.

MISCELLANEOUS

The Vietnam era M1 helmet was a very versatile piece of equipment. The iconic helmet had been the standard helmet since the beginning of World War II. The helmet consisted of a fiberglass liner with riveted web suspension. A metal cover commonly called a "steel pot" fit snuggly over the liner. Its primary purpose was protecting the head from injury from shrapnel—it was not designed to stop a direct hit from a bullet. During the Vietnam era, the steel pot was fitted with a cloth camouflage cover. The cloth cover was reversible, green on one side and brown on the other. It was held in place by flaps that folded between the steel pot and the plastic liner and by an elastic band that fit snuggly around the helmet just above the flare at its bottom. Grunts frequently employed this band to carry small items on their helmet. The most common items were the small plastic bottles of insect repellent and cigarettes. The chinstrap never was worn. It was always stowed on the back base of the helmet. The helmet could be used as an emergency entrenching tool or as a weapon in hand-to-hand combat. However the steel pot's most common unsanctioned function was as a washbasin for shaving and bathing. It also made an excellent cooking pot.

I have a fond memory of cooking stew in a helmet once. On that occasion, members of first platoon's first squad filled a steel pot with several cans of C–Rations: chicken noodle soup, ham slices, beans and wieners, etc. We added some wild peppers and sliced bamboo shoots. We seasoned it with salt, pepper, and Tabasco® hot pepper sauce (most of us carried a small bottle of Tabasco® sauce that we received from home in a care package). Then we cooked the stew over a fire fueled with C-4 plastic explosive. C-4 burned very hot. So we had to stir it constantly. The plastic spoons from our C-rations were too flimsy for this, so we employed one grunts' hunting knife. That improvised stew proved to be quiet tasty.

The company had a beer and soda fund. We did not have any means to cool the drinks in the bush. Nevertheless we periodically received a few cases of beverages with our resupply. The usual allotment was one 3.2 beer[267] and two sodas per man. The beer had low alcohol content and tasted terrible hot. I always traded my beer for sodas. I usually managed to acquire at least two sodas in the exchange for one beer. Dr. Pepper was my favorite beverage. It was the only soda that still tasted good when it was hot. All of the beverages came in cans that required a "church-key" style civilian can opener.

One other item frequently included in a resupply was a larger cardboard box with the block letters SP printed on the side, which stood for Sundry Pack. Its contents included cartons of cigarettes, stationary and envelopes, bootlaces, toothpaste and toothbrushes, shaving gear, candy, and paperback books. These articles were rationed to each soldier. I used my allocation of cigarettes from the SP pack and those issued in C-ration boxes to barter for other items I needed. Bootlaces not only were used in combat boots, they also tied

---

267. The beer was only 3.2% alcohol by weight. On average, most beer contains 4.5% alcohol.

up poncho hooches in the bush. They also had numerous other practical uses in the bush.

DEROS

The weather in Nam was hot in September. The day I turned in my M-16, it was 106º F in the shade. I remember that because there was a thermostat hanging in the shed where I cleaned the weapon and completed the customary army paperwork. At 8:30 p.m. on September 5, 1971, I sat down and wrote out a brief account of my tour in Vietnam. I ended the narrative with these words:

> In eight more days I'll be home and my time in this "Hell["] will have ended. I thank God for my life and end with a statement often engraved on the lighters of Viet veterans:
> "Life has sweet taste the protected will never know."
> "Freedom has a special meaning for those who have fought for it."
> Various versions of the two exist but they all say Deros brings indescribable relief and may it be: the beginning

I finished processing out of the division the following day. That night three of us who were leaving Vietnam walked over to the enlisted men's club at SERTS. We all three got drunk, and I mean DRUNK. I have never been so intoxicated in all my life. The whole purpose for going to SERTS was to haze the new cherries who were beginning their tour of duty in Vietnam. However we were so smashed that we foolishly walked back to our battalion area along the green line where these cherries pulled their first in-country guard duty. They were frightened and trigger-happy. Fortunately, our celebratory inebriation clearly identified us as friendly.

The next morning I left Camp Evans, never to return. I traveled by truck to Phu Bai. From there I boarded an Air Force C-130 for Cam Ranh Bay. Upon arrival at Cam Ranh Bay I made my way to the 22nd Replacement Battalion depot. With each progressive step, the reality that I was leaving Vietnam raised the tension I felt.

After reporting to the 22nd Replacement Battalion, the first item of business was a mandatory drug test. Soldiers who failed the test could not return Stateside until they completed a detoxification protocol. I had never touched any drugs, not even marijuana.[268] Therefore this presented no problem for me.

Formations were held periodically throughout the day. During these formations, the manifest for the next flight stateside was announced. Those whose names were called left the formation to complete their processing out of country before boarding the airplane. So if you missed these formations, you might miss your flight home. On the down side, after the people in a manifest separated from the formation, other people were selected for various work details, such as picking up trash.

At one formation I was chosen for a detail that was going to rake the gravel in an area of the replacement depot. The NCO in charge marched us in a column of twos to the supply shed to pick up rakes. As he bellowed the preparatory command, "Column right!" the guy next to me hissed, "Double time!" As the NCO unsuspectingly thundered, "March!" the two of us sprinted forward and disappeared around a nearby building. Since the detail was selected by position in the formation and not by name, the NCO was unable to identify the

---

268. Marijuana usage was widespread among the military in Vietnam. The Department of Defense estimated in 1971 that 51% of military personnel in Vietnam had smoked marijuana. Perhaps as many as 30% of them had experimented with psychedelic drugs such as LSD.

culprits who escaped his charge. If he left the remaining troops in his detail, he would have no one to rake. He could not identify us by appearance; we were dressed in green jungle fatigues like everyone else.

Finally, on the afternoon of September 13, 1971, while standing in formation I heard the NCO calling names for the manifest of the next freedom bird[269] say, "Matthews, Edwin L." A bolt of excitement shot through my body as I caught my name. At last I was leaving Vietnam forever! Later that afternoon I walked across the tarmac and climbed the stairs to board the freedom bird that would take me back to the world, a Flying Tigers Airline Douglas DC-8. Inside the cabin the only sounds were the crew preparing for takeoff. As a stewardess shut the cabin door I felt a certain sense of relief. The relief subsided, replaced by a further anxiety as the airplane taxied out onto the taxiway. My emotions were not unique. The airplane came to a stop at one end of the airstrip. The big jet slowly turned to face the runway. Many hearts beat faster with anticipation as the pilot revved up the aircraft's four engines. Passengers held their breath as the aircraft raced down the runway, vibrating noisily. Suddenly the din changed to a still tranquility as the airplane lifted into the air. An earsplitting spontaneous cheer reverberated throughout the cabin. A couple of stewardesses scolded about childish behavior, but they could never understand the joy unleashed when the plane's tires separated from the concrete. For the over 200 passengers inside, the war was over. We were going home! Good bye, Vietnam! Hello, U S of A!

---

269. "Freedom bird" was the slang designation for the airplane carrying those troops who had completed their tour of duty back to the United States.

HOME

Marriage, ETS, and College

FORT LEWIS AND HOME

We stopped briefly in Japan to refuel. Inside the terminal I purchased a ticket to fly from Seattle-Tacoma International Airport to Atlanta, Georgia. Due to the International Date Line, the purchase date was later than the flight date. Afterwards I joked that today I bought a ticket for a flight on yesterday, but made the flight on time.

The DC-8 entered U. S. airspace near San Francisco, California. The sunrise was beginning to burn away the morning mist, revealing the American coastline. I put my Instamatic camera against the window opening and snapped a picture of San Francisco Bay. After crossing the shoreline the airplane turned north and flew to McChord Air Force Base in Washington state. Buses transferred all Army personnel to Fort Lewis for further processing. One of the first items on the agenda was being fitted for Class A dress green uniforms. Short-sleeved khakis were the standard year round dress uniform in Vietnam. However, the season when khakis were approved to be worn in the continental United States had passed. With the advent of cooler weather, only dress greens were authorized. We could not leave Fort Lewis until we donned these regulation stateside uniforms. Each man was measured and his measurements recorded so the tailors could alter the uniform while we completed other segments of processing.

After finishing measurements at the tailor, we strolled to the special mess hall that fed troops returning from Vietnam. The interior was set up like a civilian restaurant instead of an army mess hall. The tables seated four and each meal was individually prepared. Every returnee was entitled to one steak dinner. The steak was real beef prepared to your individual taste. I had a steak, medium rare, a baked potato with butter and sour cream, tossed salad with Thousand Island dressing, a dinner roll with butter, and chocolate cake.

After lunch we were supposed to be paid. But the paymaster informed us that we had to be in our dress green uniforms to receive pay. So far the only components of that uniform that we had received were the tan poplin shirt and black tie. Therefore we waited. Finally the base commander issued a special order making a special dress uniform for the ad hoc unit of returning veterans. It combined the poplin shirt and black tie with the summer khaki trousers. Why he didn't just authorize the summer khaki uniform that everyone was wearing, I cannot say. It seems to have been another case covered by the proverbial "army way."[270] We were just happy to be paid. We would need real American currency for travel expenses in the United States.

Flushing toilets had been a forgotten memory. Now they were a curiosity. In the bush, when nature called, a grunt grabbed an entrenching tool and dug a small hole. After completing his business, he

---

270. The proverb reads: "There is the right way. There is the wrong way. Then there is the army way."

refilled the hole and covered his waste. So flushing the toilet several times just to watch it operate was not unusual. Showers and sinks likewise would provide fascinating thrill. In the bush, rain or rivers had provided these amenities.

The hour was late when our in-country processing was completed. Several of us went in together for a taxi to take us to Sea-Tac (Seattle-Tacoma International) Airport. I had just enough time at Sea-Tac to eat a light supper before boarding the airplane. I flew Braniff International flight 189 to Dallas, Texas. In Dallas I would catch a Delta Airlines flight to Atlanta and home. However, I had a long layover in Dallas. It was late and most services in the airport were closed. I was too impatient to sleep. Fortunately, another soldier on the same Delta flight as me was sitting in the desolate waiting area. He was heading to Canton, Georgia. We partnered up for the trip to Georgia. We passed the long hours of waiting in conversation about finally making it home and planning the final leg of our journey—the trip from the Atlanta airport to our respective homes. He had a friend who volunteered to pick him up at the airport in Atlanta. So he invited me to ride with them, promising to drop me off in my parents' driveway. I gladly accepted the offer. I had not been able to telephone home yet. Now I decided to surprise Daddy and Momma. I calculated that I would arrive home about the time Mother was cooking breakfast.

Delta Airlines flight 256 to Atlanta left Dallas at 2:30 a.m., Tuesday, September 14, 1971. It was still dark when the airplane landed. I was sitting by a window, searching for any familiar sight. The green "Atlanta" sign on the airport terminal was the prettiest sight in the world. It signified I was nearly home. Furthermore, I had a 30-day leave in which to enjoy being home.

Just as he promised, my travel companion's friend picked us up at the airport. It was still dark as he pulled onto I-285, the perimeter interstate highway that circles metropolitan Atlanta. I was unaccustomed to riding in the back seat of a car driven by an American teenager. The trip on I-285 was more terrifying than a hot LZ north of Khe Sanh. The speed limit was 70 mile per hour, but he drove at least 10 to 15 mph faster. I had ridden last on a highway in the back of an army deuce-and-a-half truck and this speed seemed like 800 miles per hour in my mind. A steady stream of bright headlights from oncoming traffic flashed into my eyes, destroying my night vision. Our driver whipped in and out of red taillights. To my wartime senses, they all blurred into incomprehensible streaks of light. As we traveled around the perimeter, sunrise dawned. The growing sunlight allowed me to differentiate the sources of the lights as cars. Simultaneously, traffic increased as rush hour commenced. Finally, the car slowed as we veered onto the exit ramp to South Cobb Drive. Its recognizable sights seemed like a familiar fantasy. *Was I really coming home? Was the war really over for me?* We stopped in the left turn lane at Jones Shaw Road (now Windy Hill Road) and waited for the traffic light to change to green. My heart was pounding with excitement. The feeling almost compared to that of going into a hot LZ, but at the same time was distinctly different. We turned right and rode down Wynona Drive. At long last our car turned onto Okemah Trail and stopped at the first house on the left, a red brick house with white wood trim. My 1969 green Chevrolet Nova was parked in the driveway. I was home! I stepped out of the car and walked up the driveway. My heart was beating so hard now that I knew it was going to explode!

Mother was frying bacon for breakfast. She held the iron skillet in one hand with a potholder. The pan was about an inch above the stove. With a fork in her other hand she prodded the bacon strips. The kitchen door was directly behind her. When I opened the door she turned to see who was coming into the house. Recognizing me immediately, she dropped the skillet and screamed! Then she stood frozen, tears running down her cheeks. I went over and hugged her.

Daddy came scrambling into the kitchen to check about Mother's scream. When he saw me a big smile erupted on his face and tears ran down his check. Startled by their parents' inexplicable noises, Reita and Lisa slipped into the hall to investigate. Seeing Daddy and Momma hugging me, they raced down the hallway to join the family celebration. There only was one more person I wanted to see, Pamela. Until I could hold her in my arms and look into her blue eyes, I would not be completely home. The memory of her and our times together had gotten me through the darkest of times while I was away. I needed to see her to bring some closure to Vietnam.

* * *

My body retained a foul odor after Nam. Bathing could not remove the stench left in the body by the experience. Mother followed me around the house spraying Lysol® disinfectant everywhere I went. Whenever I ran bathwater, she poured liquid Lysol® disinfectant into the tub. My sister Reita worried that all that Lysol® might be harmful to her big brother. In time the stink faded. I soon smelled just like the people who had never been to Vietnam. But all the Lysol® in the world could never wash away the experience of Nam. Odor is a powerful trigger to one's memory. Even today the smell of diesel fuel rekindles memories from that far away time.

MARRIAGE

I didn't see Pamela the morning I got home. It was Tuesday and she had to go to work at Chubb-Pacific Indemnity Insurance in midtown Atlanta. She only had been employed since mid-June and so taking time off was limited. Our wedding was planned for two weeks later and she already had scheduled two days off for the wedding. I was too nervous to go into Atlanta for lunch. So we didn't see each other until later that evening. The pinnacle of my homecoming came when I held her petite body in my arms for the first time since April. I looked into her bright blue eyes and pressed my lips against her glistening lips. I was home for real. Vietnam was behind me. Pamela was before me. God truly is good. The life God gave me at birth was good. And the life God spared in Vietnam was—and is—especially good. Until one has seen just how fragile life is, and how quickly it can be snuffed out, one cannot truly appreciate how good life is.

Pamela and I were wed on Thursday, September 30, 1971, at Green Acres Baptist Church in Smyrna, Georgia. Originally we planned for a large formal wedding, but difficulties in making plans while I was in Vietnam—everything had to been done by mail—eventually led us to limiting the guest list to immediate family and one friend each. The limitation was fortuitous. I could not have coped with a large crowd staring at my back. Like many Vietnam combat veterans, I suffered from PTSD.[271] At the time, the disorder was commonly called Vietnam War Syndrome by mental health specialists.[272]

---

271. "Post-traumatic stress disorder (*PTSD*) is a mental health condition that's triggered by a terrifying event —either experiencing it or witnessing it. Symptoms may include flashbacks, nightmares and severe anxiety, as well as uncontrollable thoughts about the event." Quoted from the Mayo Clinic website https://www.mayoclinic.org/diseases-conditions/post-traumatic-stress-disorder/symptoms-causes/syc-20355967. Accessed November 25, 2019.

272. Except in severe cases PTSD sometimes went undiagnosed and untreated. Although I never received formal treatment, conversations with other veterans eventually proved liberating. My calling as a pastor often provided numerous opportunities for such exchanges. In the early 1990s another Vietnam veteran and I formed a support group where vets could discuss their experiences in and after

I was dressed in my Class A dress greens, the same ones that delayed processing at Fort Lewis. She wore the gorgeous white formal wedding dress that she purchased for the large wedding. My father was best man and Pamela's sister-in-law Cathy Strickland was maid of honor.

Following the ceremony we drove across the street to the home of Pamela's parents for a wedding reception. During the wedding reception, my best friend Dale Cheney, my sister Reita, and my new sister-in-law Cathy decorated my dark green Nova with white water-based paint. They also tied tin cans to the rear bumper. For a moment, my temper flared when I discovered this, but Pamela calmed me down. After the reception we headed to Chattanooga, Tennessee, for our honeymoon. We paused around the corner from the house just long enough for me to remove the cans. We stopped at a Bar-B-Que restaurant in Cartersville, Georgia, for dinner. Then we drove on to Chattanooga. My wedding band was too big. I kept my hand in a fist so it would not slide off. After Vietnam, the idea of being a newlywed was euphoric.

Reita gave Pamela and me reservations at a Holiday Inn motel for our wedding present. Words cannot describe my emotions when I drove up to that motel on the south side of Chattanooga. I was behind the wheel of MY car. MY wife sat beside me. MY new life in the world was taking shape.

We visited Rock City and Ruby Falls, two popular tourist attractions, on Friday. On Saturday we took it easy. We watched television and dosed off at separate times. I watched Georgia Tech defeat Clemson 24-14 in college football. She watched the San Francisco

---

Vietnam. The church where I was pastor hosted the group. At its peak this group had 20 to 30 regulars. As the Veterans Administration developed better treatment, our group assimilated into VA groups.

Giants edge the Pittsburgh Pirates 5-4 in the National League baseball playoff game.

Pamela and I came back to her parents' house after the honeymoon. Walking into their living room that Sunday afternoon with our suitcases was one of the most awkward moments of my life. We had decided that Pamela would remain with them while I finished my time on active duty. I was scheduled to ETS on March 31, 1972. ETS — Estimated Time of Separation — was the date a soldier left active duty. During the Vietnam era, draftees normally spent two years on active duty, followed by four years in the reserves. In late 1971, the withdrawal from Vietnam was causing the Army to give soldiers an early out. So we knew that almost assuredly I would be out of the Army long before that date. Pamela had a good job at Chubb. If she quit it she might not find another one a few months later. Federal law guaranteed that I could return to my civilian job, but it was part-time with Rich's Department Stores. Therefore we determined that our wisest course was for her to keep her job and remain in Smyrna and pray that I got out long before March 1972.

BACK WITH THE 4TH INFANTRY DIVISION

After my leave I reported to the Fourth Infantry Division at Fort Carson, Colorado. The division was now mechanized and this time I was assigned to Company C, 1st Battalion, 11th Infantry (Mechanized). The entire company consisted of Jack Cole, Allen Elick, Ron Reeves, and me. A four-man infantry company is pointless. We were too undermanned to perform any military role. Sufficient personnel were not even available for a detail to police the company area. However the four of us kept it clean, thereby evading the need for such menial tasks.

The company would eventually be brought up to its proper

strength, but it would be weeks before more replacements arrived. In the meanwhile, the army seemed not to know what to do with the four of us. We were assigned to an M-113 APC (Armored Personnel Carrier) since we were supposed to be a mechanized infantry unit. However it never left the motor pool while I was assigned to the company. A training exercise for one APC with a commander, a driver, and two infantrymen did not exist.

I eventually was detached to train in the battalion armory. I spent several days learning to work on M-16s and other light weapons. This clearly was an effort to occupy my time during duty hours. Mostly I passed the time with army red tape. I had to process into the division, a boring progression of reporting to various stations to fill out required paperwork. Later I had to return to the same place and process out of the division. On some occasions I would process in during the morning and process out that same afternoon. Sometimes a space of a few days separated the processing in from the processing out. Frequently I processed out of the division at one station before I processed into the division at another station.

The case of my TA-50-901 gear was a classic example of rigidity and absurdity in army bureaucracy at the time. TA-50 gear comprised the field equipment an infantry soldier required to carry out combat operations. As an infantryman assigned to the Fourth Division (Mechanized), I was issued the standard division TA-50 gear. So on Monday morning, November 15, I went to supply and picked up my gear. Captain John O. Howell, Jr. had been assigned command of Company C. I obtained from him written authorization to return the equipment early. His letter also stated, "This equipment was inspected and found to be a complete issue, and in the same condition as was issued."[273] After lunch I returned to the supply depot and laid

---

273. Paperwork dated "15 Nov. 71."

the bag of equipment on the counter. A private wearing the single chevron of an E-2 dumped the bag on the counter. His rank and new uniform suggested he likely had been in the army only three or four months. Nevertheless this neophyte paper shuffler informed me that my equipment was dirty and I would have to clean it before he could accept its return. The supply clerk's arrogance infuriated me. *I'm out of Vietnam and I still have to deal with idiot REMFs!*

The equipment was in the exact same condition as it was when issued two hours earlier. It had not been out of the bag. I don't know exactly what all else I said to him, but I did say, "There is no d*** way I am cleaning this stuff!" I was loud enough to attract the attention of a nearby Staff Sergeant. He immediately took charge of the situation. I explained the situation to him and showed him the CO's letter. While I was talking, he not only listened to my words, he also assessed me as an individual. His scrutiny of the CIB over my breast pocket and the Screaming Eagle patch on my right sleeve was obvious. Before I finished speaking he picked up my paperwork from the countertop, tore off copies for me, and told me I could go. I didn't clean the equipment. During the six weeks I spent at Fort Carson, much of the time was spent in similar frivolous exercises of military formalities.

Duty at Fort Carson also had its positive side. At the time the army post was testing new concepts to enhance the forthcoming all-volunteer army. One of these ideas was specialty battalion mess halls. These specialty messes were allocated as supplements to the regular army mess halls. Each battalion had a specialty mess hall with its own culinary specialization. They were open from lunch to midnight. No limit was placed on the number of times one ate or upon the quantity of food consumed. A soldier just signed a form each time he went through the serving line. My battalion's specialty mess was Italian food. The tables had red-checkered tablecloths and candles inserted into the mouth of a wine bottle. Overall the décor provided a pleasant atmosphere. I also frequently visited the Mexican mess hall of the adjacent battalion for its tacos.

One weekend, the four of us grunts in my company took an excursion up into the Rocky Mountains. Ron had his old Volkswagen Beetle. Unfortunately, it was tuned for sea level conditions at his home in Florida instead of the high altitudes of the Colorado Rockies. As we reached the higher elevations, the engine commenced cutting out. So we three passengers often were forced to get out and push when we started uphill. Nevertheless we visited Pikes Peak and the small town of Cripple Creek before returning back to base.

By mid-November I received an official ETS date of 1 December 1971. I would be home for Christmas. November had been abnormally mild that year in Colorado Springs. The weather changed drastically on 1 December 1971...my ETS day. A major blizzard struck that day, the day I was getting out of the army! Naturally the army saw no need to hurry. I, on the other hand, was anxious to get out of Colorado before I was snowed in. It was snowing hard and accumulation was building up fast. I was handed my DD-214 discharge paper. I scanned it and noted that my Air Medal and Bronze Star Medal were not listed. I thought about having it corrected, but decided that since changing the form would take time, I wouldn't bother. Getting home to Pamela was more important. I had copies of the orders for both medals and the snowfall was getting heavier.

Ron drove me to the airport. I walked into the terminal and a girl behind the Continental Airlines counter hollered with the news that they had one flight leaving for Denver. It would be the last plane out that day. The authorities were shutting down the airport because of the foul weather. She quickly wrote out a ticket and told me that a flight attendant on the airplane would assist me in getting to my final destination. During the short flight one of the flight attendants took appropriate information from me. When I debarked at the gate in

Denver, a ticket for Atlanta was waiting for me at the counter. Due to weather conditions—the airport at Denver also anticipated closing soon—I would need to fly first to O'Hare International Airport in Chicago, Illinois. After a brief delay in Chicago, I would transfer to an Eastern Airlines flight to Atlanta. Snow was falling in Chicago, but the precipitation caused no problems at O'Hare. The flight to Atlanta was noneventful. Pamela picked me up at the airport and drove us to our first home, an apartment in Smyrna, Georgia.

GRANNY'S DEATH

We had Christmas dinner at Daddy and Mother's house that year. Granny, Paw Paw, and Mother's sister Katherine were there. About an hour after they left to drive back to Chattanooga, the telephone rang. They had been in a serious automobile accident. Granny had been killed. Paw Paw would spend the next few months in the hospital. Katherine, only was hospitalized a week. After her discharge, she stayed with Mother until Paw Paw recovered and they were able to move back into their home in Chattanooga.

 I remember standing in the pew during Granny's funeral and being unable to cry. I loved her. Her death hurt and it troubled me that I could not grieve openly for her passing. But I had become too hardened to death in Vietnam. Granny was 75 years of age when she died. She had lived a full life. In contrast, I had seen too many 18 and 19 year old boys killed. Their lives had been cut short way too early. Therefore the tears simply would not flow.

 To this date I do not know the reason for it, but Granny had a life-long phobia of me going away to war. I remember that even when I was a child she would tell me she prayed that I never would go to war. But God in His wisdom and sovereignty did not grant her request. However, I believe He allowed her to live long enough to see me return safely from my war.

WELCOME HOME

I had wanted to attend Georgia Tech since high school. While in Vietnam I applied for admission to the Institute.[274] I had spent a year in Vietnam because I failed a French course at Kennesaw Junior College. Hence the rejection letter I received back came as no surprise. I tossed it into the trash and forgot about it. If I survived and returned home, I could decide what to do in civilian life. Surviving required my attention being focused on doing my job, not delusions about college.

I was out of the army and married. The future was now. The wife I had longed for in Vietnam was now reality. *I want to have children one day. If I'm going to have them, I must take care of them. I better take care of their mother.* The time had come for me to take responsibility for the life God granted me on that slope near Purple Heart Hill.

I returned to my job with Rich's Department Stores. Christmas was approaching and they were happy to have me back. The job provided some income, but I didn't want to spend the rest of my life on a loading dock. I planned to go back to Kennesaw Junior College and improve my academic standing. Then I could reapply to Tech with a real prospect of acceptance.

To my surprise, just before Christmas I received an invitation to orientation at Georgia Tech. When I contacted Tech, they affirmed that I was scheduled to attend orientation. At orientation I was told

---

274. The official name of Georgia Tech is the Georgia Institute of Technology. Those attended there or are associated with the school refer to it as "the Institute" and highly resent people referring to it as a university or college.

to report to the registrar's office on the first day of school if I did not receive an acceptance letter. Of course I did not—I already had received a rejection letter.

So on the first day of classes at Tech I reported to the Registrar's office. When the Registrar pulled my file he said, "Someone made a clerical error. You should not be here."

"Yes sir, but I am here. What do you plan to do?" I replied, expecting him curtly to send me home.

"Son, you are here. We can use the money. I'm going to accept you. You will be on probation initially."

He paused for a moment, stared directly into my eyes and said emphatically, "But you can't make it here at Georgia Tech."

The Registrar was going to admit me on the GI Bill, but categorically told me I could not succeed at Georgia Tech. I did not like being told I can't do something. One positive legacy from Vietnam was an inbred determination to do whatever needed to be done in order to accomplish an assigned task. If I could survive a year-long tour of duty in the jungles of Vietnam as a grunt, I surely could do what was necessary on the campus of Georgia Tech! I made the Dean's list the first three quarters. I finished with two degrees, a Bachelor of Science and a Master of Architecture.

I know that I survived Vietnam by the grace of God. Eventually I also would earn a Master of Divinity degree and a Theological Doctorate from New Orleans Baptist Theological Seminary. I would pastor Baptist churches for thirty years. After retiring from the pastorate, I served as Director for the New Orleans Seminary extension center in Columbus, Georgia. Pamela and I are still happily married to each other. We have four terrific children and seven wonderful grandchildren.

In Vietnam, the grunts dreamed of the world they left and longed to return there. But as Mark Baker astutely observed, "They never returned to the world they had left."[275] In the year they were away, their world had changed. They too had changed. In the popular image, the Vietnam veteran is a troubled antisocial misfit. Sadly, too many of these stereotype vets do exist. The war in Southeast Asia destroyed their future. All of us bear the scars of that unpopular conflict. However, in reality, most Vietnam survivors returned home and became productive members of society. But the war always will be there within us grunts. We live daily with horrific images of what we saw and experienced during our tour of duty in the Republic of Vietnam.

---

275. Mark Baker, *Nam: The Vietnam War in the Words of the Men and Women Who Fought There* (New York, New York: William Morrow and Company, 1981), p. 213.

PHOTOGRAPHS

LeBron Matthews' basic training photograph, 1970

THE FLAG WAS STILL THERE

Pamela Jean Coffman, 1969 —
this is the photograph LeBron carried in his wallet while in Vietnam

LeBron as a RTO in 1971

A gun truck from the 4th Infantry Division leads a US Army convoy through An Khe Pass, October 1970

The Rakkasan (3rd BN 187th Infantry Regt., 101st ABN Div.) display in the battalion area at Camp Evans

A UH-1H Huey lands on the Rakkasan helipad at Camp Evans— the giant torii designated the battalion's area

The cave that the North Vietnamese Army used as a supply bunker in the Falls AO as viewed from Charlie Company's position during the firefight on January 19, 1971

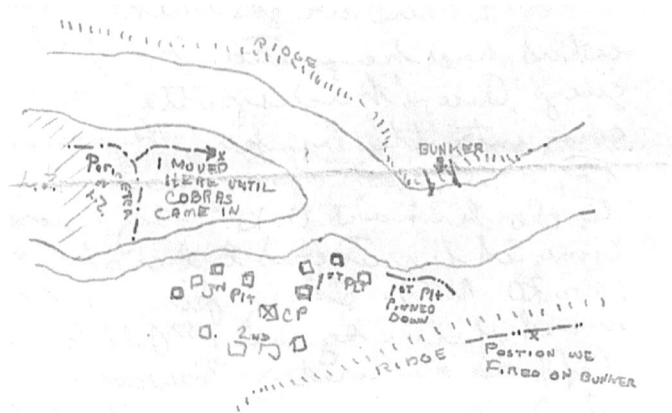

Figure 1 — My sketch of the firefight in the Falls Operation, 17 January 1971. The sketch was drawn three days after the action.[276]

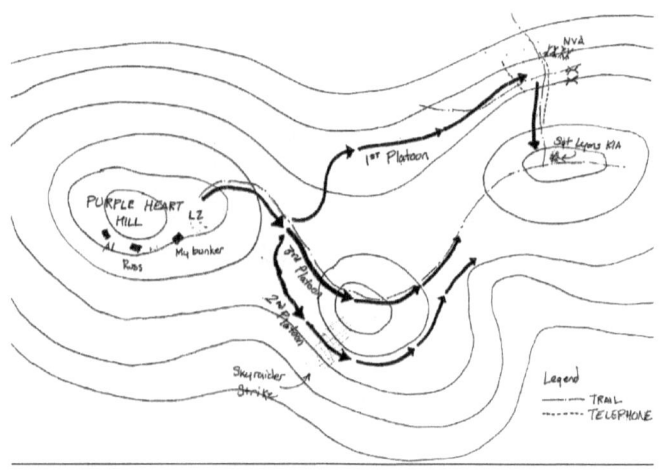

Figure 2 — Sketch of attack by Co C, 3/187 Inf, 101 ABN Div, on February 20, 1971 (not to scale)[277]

---

276. Letter to my parents 20 January 1971

277. Drawn by the author, Friday, March 23, 2018. Based upon discussions with Russ Kagy and Alan Davies, February 8-10, 2018.

Members of 1st platoon, Company C, 3/187 Infantry, 101st ABN Div. after the Falls operation—1st row (left to right): Carl (Snuffy) Booth, Joe Sej, Hector (Recon) Torres, Cecil Beach, Larry Klag. 2nd row Kevin Owens, Alan (Uncle Al) Davies, George (George of the Jungle) Bloom, Russ Kagy. back row: unknown, LeBron [directly behind Kevin & Alan], unknown, Sgt. Jim (Egg) Eggleston [with pipe], unknown.

Bunker occupied by LeBron, Hector (Recon) Torres, Larry Klag, and Joe Sej on Purple Heart Hill during the bombardment on February 21, 1971

The view looking toward the Rock Pile from LeBron's bunker on Purple Heart Hill

THE FLAG WAS STILL THERE

Flag over Purple Heart Hill after the bombardment on February 21

Members of 1st platoon, Company C, 3/187 Infantry, at Camp Carroll after Purple Heart Hill – 1st row (left to right) me [wearing a NVA belt], Larry Klag, Hector (Recon) Torres. Back row. Joe Sej, (Cowboy), (Jenks), unknown, Carl (Snuffy) Booth.

Company C 3/187 Infantry linking up with the M48A3 tanks of 1/77 Armor during Dewey Canyon II/Lam Son 719

Captain Metcalf (center of photo) inspecting equipment the morning after Typhoon Harriet, July 1971

THE FLAG WAS STILL THERE

LeBron's 21st birthday—Geary Mortimer (left) and LeBron (right) holding a C-Ration pound cake with a cigarette substituted for the candle

ABBREVIATIONS, ACRONYMS, AND ARMY SLANG

Some acronyms were pronounced as words. The pronunciation of these terms is given in brackets.

a.k.a.—Also known as

ABN—Airborne

AFVN—American Forces Vietnam Network. The United States military radio and television network in Vietnam.

AIT—Advanced Individual Training, the army's training course for a recruit's MOS (see below).

AK-47—A 7.62mm assault rifle. The Kalashnikov AK-47 was the primary infantry weapon of the Communist VC and NVA troops.

AO—Area of Operations, the term designated the region in which a unit carried out its assignment.

APC—Armored Personnel Carrier. In Vietnam, the M113A1 was the mainstay of mechanized infantry.

ARA—Aerial Rocket Artillery, official name for some helicopter gunships.

ARC LIGHT—Code name for a B-52 bomb strike in South Vietnam or Laos.

ARVN [r-vun]—Army of the Republic of Viet Nam, the South Vietnamese regular army.

B-52—The Boeing B-52 Stratofortress, known as "Buff" to Air Force personnel, was an eight engine jet bomber. These bombers routinely bombed from high altitudes, 30,000 feet or higher.

Beaucoup—Slang expression for "many" or "much." The word originated in the French language and commonly was employed in communicating with the local Vietnamese population.

Bird—Slang term for a helicopter.

BN—Battalion

Boom-boom girl—Slang for a prostitute

Boonies—Slang for the jungle and any area outside a base camp. It also was called "*the bush.*"

Brown boot army—Slang for pre-1956 army. The Army began changing from brown leather boots to black leather boots in 1956. Career soldiers were known as "lifers." Lifers who began service before 1956 were said to have come out of the brown boot army.

Bush—Slang for the jungle and any area outside a base camp. It also was called "*the boonies.*"

C&C—Command and Control helicopter, used by the commanding officer to supervise his unit's movements

C-130 Hercules—A four engine cargo airplane built by Lockheed Aircraft in Marietta, Georgia. The C-130 was the principle Air Force cargo aircraft in Vietnam.

C-7 Caribou—A small twin engine cargo airplane.

CA—The military abbreviation for combat assault. The acronym was applied to any helicopter movement into hostile territory. If the incoming troops received hostile fire, the assault was categorized as a "hot" LZ (Landing Zone). If there was no enemy contact it was said to be a "cold" LZ.

Capt.—Captain

Cherry—A slang term for replacements, a soldier just in Vietnam from the United States.

Chopper—Slang term for a helicopter.

CIB—Combat Infantry Badge, a badge award to trained infantry MOS soldiers who had served in an infantry unit operating in a combat zone for an extended period or who had come under enemy fire while serving in an infantry unit. Among infantry soldiers in

Vietnam it was a high prized award in that it designated the wearer as a member of the fraternity of combat experienced infantrymen.

Claymore Mine—The M18A1 "Claymore" APERS Mine was an electronically detonated anti-personnel mine developed for perimeter defense. The mine was set up above ground and could be set up and removed easily.

CO—Commanding Officer

Col.—Colonel

Connex—An intermodal container is a standardized reusable steel box that is used in shipping and storing freight.

CONUS [co-nus]—The military acronym for Continental United States.

CP—Command Post, a unit's command and communication center. The acronym designated the personnel and/or their geographic location. At the company level, in the field it usually accommodated the commanding officer and two radio men. Frequently it also contained an artillery FO (Forward Observer) whose job was directing artillery fire to support the company.

CSM—Command Sergeant Major

D. I.—Drill Instructor

Dee dee, dee dee mao—Slang meaning "to leave"

Delta Tango—Designated Targets; pre-selected artillery targets located in daylight for use at night

DEROS [dē-rōs]—Date of Estimated Return from Overseas, the date that a soldier was scheduled to rotate out of Vietnam and return stateside.

Dink, Dinks—3rd of 187th's preferred derogatory slang term for the NVA and VC. Elsewhere *Gook* was a common slang designation.

Dinky-dao—Slang (Americanized Vietnamese) for "crazy"

Div.—Division

DMZ—Demilitarized Zone; The DMZ was established by the 1954 Geneva Conference. It was a five mile wide buffer zone along the 17th parallel that separated North Vietnam and South Vietnam.[278]

Doc—Universal nickname for all medics

DoD—Department of Defense

Drag—The last soldier in a column (usually in single file). His responsibility was rear security, watching the trail over which the column has just moved for enemy units who might be following the column.

Dust-Off—A term for medical evacuation, Dust-Off units, *a.k.a.*

---

278. *The Encyclopedia of the Vietnam War*, 2000 ed., s.v. "Demilitarized Zone (DMZ)" by Brent Langhals, p. 97.

Medevac, employed Huey helicopters fitted as air ambulance to evacuate wounded soldiers off the battlefield.

ETS—Estimated Time of Separation, the date a soldier left active duty. During the Vietnam era all enlistments were for six years. Draftees normally spent two years on active duty, followed by four years in the Army Reserve. Theoretically the reserve time was divided into two years active and two years inactive.

FAC—Forward Air Controller

FNG—F***ing New Guy, a popular slang term for replacements, a soldier just in Vietnam from the United States.

FO [sometimes: fō]—Forward Observer, a soldier from an artillery battery or an Air Force squadron who was assigned to accompany an infantry unit in order to direct artillery fire or bombing missions in support of the infantry unit.

Frag—(1) as a noun, a M26 fragmentation hand grenade; (2) as a verb, "to throw a grenade," especially as in "Fragging" (see below).

Fragging—Tossing a hand grenade at an individual—usually a senior NCO (Noncommissioned Officer) or an Officer—or an area where he frequently was known to be. Although it gained more notoriety in Vietnam, it was equally common in every war.

Freedom Bird—American slang for the airplane that carried soldiers back to the United States after their tour of duty in Vietnam was completed.

FSB—Fire Support Base (commonly call a firebase), an artillery position.

FTA—"F*** The Army" was a common expression of discontent used by draftees serving in the army in Vietnam.

G.I.—Government Issue, a slang expression for an American soldier: it was made popular during World War II and continued to be used afterwards.

Gen.—General

Grunt—A popular slang name for an infantryman in Vietnam. It was allegedly derived from the sound occasioned by an infantryman lifting up his rucksack.

HHC—Headquarters and Headquarters Company

Hooch—Soldier's slang for living quarters in Vietnam. The term had a wide range of application, from temporary shelters constructed in the bush to permanent barracks in permanent posts.

Huey—Slang for the Bell UH-1H Iroquois helicopter. The term was applied to all versions, but by 1970 virtually all of the UH-1s in Vietnam had been upgraded to the H series.

Indian country—Army slang for areas controlled by the North Vietnamese.

Inf.—Infantry

KIA—Killed in Action

KP—Kitchen Patrol, consisted of duty in the unit's mess hall. Tasks ranged from washing dishes, pots and pans, to mopping the floor, to peeling potatoes, to serving meals.

Land of the big PX—"land of the big post exchange," American slang for the United States of America. The PX, or post exchange, were retail stores on military installations. These stores sold items from civilian clothing to major appliances. Hence the slang expression emphasized the ready availability of retail items in America.

LCM—Landing Craft, Mechanized

Lifer—Derogatory slang term for career soldiers.

LRRP [lurp]—Long Range Reconnaissance Patrol

LT [el-tee]—Slang for Lieutenant, taken from the abbreviation Lt.

Lt.—Lieutenant

Lt. Col.—Lieutenant Colonel

LZ—Landing Zone, an open area accessible to helicopters that carried supplies and transported troops.

M113A1—A fully tracked armored personnel carrier, known as an APC.

M-14—A 7.62mm rifle with selective semiautomatic or automatic fire. The M-14 was the primary battle rifle of the U. S. Army at the beginning of the Vietnam War. Many units in Europe still carried the M-14 in 1970.

M-16A1—A 5.56mm assault rifle with selective semiautomatic or automatic fire. The M-16 was the primary American infantry rifle in Vietnam.

M-203—A 40mm grenade launcher mounted underneath the barrel of a M-16. The M-203 replaced the M-79 in 1970.

M48—The M48 Patton tank with its 90mm gun was the primary United States Army tank in Vietnam.

M-60—A 7.62 machine gun. It could fire up to 600 rounds per minute.

M-72—A 66mm disposable rocket launcher. The M-72 was designed for anti-tank use but in Vietnam was used primarily against enemy bunkers.

M-79—A 40mm grenade launcher.

MACV [MACK-vē]—Military Assistance Command, Vietnam. MACV was the highest United States military headquarters in Vietnam.

Maj.—Major

Mechanical Ambush—An American defensive device; these innovative booby traps usually were assembled with Claymore mines and/

or grenades. They commonly were detonated by a trip wire.

Medevac—An abbreviated term for medical evacuation. Medevac, *a.k.a.* Dust-Off, employed Huey helicopters fitted as air ambulances to evacuate wounded soldiers off the battlefield.

MIA—Missing in Action

MOS—Military Occupational Specialty, the classification for job a soldier performed in the army.

MP—Military Police

MPC—Military Payment Certificate, brightly colored paper money issued in denominations from five cents to twenty dollars. MPC was the legal tender on all American military bases in Vietnam during the war.

Nam—Slang for Vietnam

NCO—Noncommissioned Officer

NDP—Night Defensive Position, usually a series of foxholes arranged to provide fire from a perimeter of 360°.

Numbah one—Number 1, slang for very best

Numbah ten—Number 10, slang for very worst

NVA—North Vietnamese Army, the regular army of North Vietnam. Also known as PAVN, People's Army of Vietnam

OCS—Officer Candidate School, the U. S. Army course for training enlisted men to become commissioned officers.

OP—Observation Post, a small position set up as an early warning for its parent unit's defense.

P-38—A small portable can opener

P.X.—Post Exchange, a retail store operated on a military installation.

PF—Popular Forces; also Regional Forces: The South Vietnamese equivalent to the United States National Guard. The PF/RF comprised 50% of the ARVN force.

POW—Prisoner of War

PRC-25 [prick 25]—The U. S. Army tactical field radio, a transistorized FM receiver/transmitter used on the platoon and company level. Weight with battery was 26 pounds.

PRC-35 [prick 35]—Essentially the same radio as the PRC-25, the PRC-35 scrambled the radio signal. A decoding device was necessary to understand the scrambled message. The device was routinely reprogramed. Weight with battery was 35 pounds.

PSP—The Pierced (or perforated) Steel Planks that engineers use to lay temporary runways. The planks were 10 feet long and 15 inches wide with rows of hole punched lengthwise.

PT—Physical Training

PZ—Helicopter pick-up zone. Enlisted men rarely distinguished between a LZ and a PZ, referring to both as a LZ.

QL—Quoc-Lo, Vietnamese for a National Highway.

R&R—Rest-and-recuperation vacation: each soldier was entitled to a seven-day R&R during a one year tour of duty in Vietnam. Out-of-country R&R could be taken in Australia, Hong Kong, Bangkok, and Hawaii.

REMF [rĕm-fah]—Rear Echelon Mother F***er, a slang term grunts used for support personnel and other soldiers who remained safe in base camps.

RIF [rĭ-fah]—Reconnaissance In Force. The acronym was the Vietnam era military terminology for a combat patrol. In Vietnam most all infantry patrols commonly were referred to as a "RIF," whether the unit was as small as a squad or as large as a company or battalion.

ROTC—Reserve Officer Training Corps, provides financial assistance for college in exchange for military service after graduation.

RPG—Rocket Propelled Grenade, a standard anti-tank weapon of the NVA and VC.

RTO—Radio and Telephone Operator. In Vietnam an Infantry RTO carried a radio.

RVN—Republic of Viet Nam, the official name for South Vietnam.

SERTS [sérts]—Screaming Eagle Replacement Training Section, the 101st Airborne Division's instruction unit for training and orientation of replacement troops coming into the division.

Sgt.—Sergeant

Short—Slang expression meaning only a brief time was left on one's tour of duty in Vietnam.

Short-timer—A soldier who was "short." At 99 days before DEROS soldiers frequently were dubbed a "two digit midget." Usually around 90 days they started keeping a *short-timer's calendar* on which they marked off each day left in Nam.

Slick—Slang term for a Bell UH-1 Iroquois helicopter, more commonly called a Huey

SOP—Standard Operating Procedure.

Spec—Specialist; Specialist Fourth Class, a Spec 4, was an E-4 rank in the army, the equivalent of corporal. During the Vietnam War the rank of corporal was rarely used. It denoted a command responsibility, whereas the Specialist classification did not. This enabled the military to promote a PFC without giving him a command position. At the same time it did not prohibit Specialist from taking command roles. Within the infantry, a fire team theoretically was commanded by a corporal. In Vietnam by 1970 fire teams universally were commanded by a Spec 4. Squad leaders were supposed to be commanded by Sergeants, but a Spec 4 squad leader was not uncommon.

Stand-down—A period of time in which an infantry unit was brought into a secure rear area for rest, refitting, and training.

The world—Soldiers' slang for the United States. It emphasized the stark contrast between the life of peace in America and life of violence in Vietnam.

TOC—Tactical Operations Center

Top—A slang term for a company's first sergeant.

Torii—The gateway to a Japanese Shinto temple. It has two posts and two crosspieces. It was emblem of the 187th Infantry Regiment and symbolized the unit as a gateway of honor.

Tube artillery—Cannons

USO—United Service Organizations, a nonprofit organization that provides various services and live entertainment to American Armed Forces and their families.

VC—Viet Cong, South Vietnamese guerrilla insurgents.

WIA—Wounded in Action.

SELECTED BIBLIOGRAPHY

MILITARY RECORDS

101st Airborne Division (Airmobile), Final Report, Airmobile Operations in Support of operation Lamson 719, 8 February—6 April 1971, volume II (1 may 1971).

Operational Reports—Lessons Learned, 101st Airborne Division (Airmobile), Period Ending 31 October 1971.

Recommendation for the Award of the Presidential Unit Citation, HQ, 1st Infantry Brigade, 5th Infantry Division (Mech), 18 June 1971.

Roster for Company C, 3rd Battalion 187th Infantry Regiment for January 1971

BOOKS

Ambrose, Stephen E. *Band of Brothers* (New York, New York: Simon & Schuster, 2001)

_____. *D-Day June 6, 1944: The Climatic Battle of World War II* (New York, New York: Simon & Schuster, 1994)

Arques, Antonio, *Grunt: A Pictorial Report on the US Infantry's Gear and Life During the Vietnam War 1965-1975* (Madrid [Spain]: Andrea Press, 2014)

Baker, Mark. *Nam: The Vietnam War in the Words of the Men and Women Who Fought There* (New York, New York: William Morrow and Company, 1981)

Boccia, Frank. *The Crouching Beast: A United States Army Lieutenant's Account of the Battle for Hamburger Hill, May 1969* (Jefferson, North Carolina: McFarland & Company, 2013)

Bonds, Ray, ed. *The Vietnam War: The Illustrated History of the Conflict in Southeast Asia* (New York, New York: Crown Publishers, 1979)

Bowden, Mark. *Hué 1968: A Turning Point of the American War in Vietnam* (New York: Atlantic Monthly Press, 2017)

Cosmas, Graham A. *MACV: The Joint Command in the Years of Withdrawal, 1968-1973*, United States Army in Vietnam (Washington, D.C.: Center of Military History United States Army, 2006)

Dustan, Simon. *Vietnam Choppers: Helicopters in Battle 1950-1975*, rev. ed. (n.p.: Osprey Publishing Ltd., 2003)

Eshel, D., ed. *M-48/60 Patton Main Battle Tank*, War Data (Hod Hasharon, Israel: Eshel Dramit, 1979)

Garland, Albert N., ed. *Infantry in Vietnam* (Nashville, Tennessee: The Battery Press, 1982)

Hammel, Eric. *Khe Sanh: Siege in the Clouds, An Oral History* (Philadelphia: Casemate, 1989)

Karnow, Stanley. *Vietnam: A History* (New York, New York: Penguin Books, 1997)

Kelley, Michael P. *Where We Were in Vietnam: A Comprehensive Guide to the Firebases, Military Installations and Naval Vessels of the Vietnam War* (Central Point, Oregon: Hellgate Press, 2002)

Moïse. Edwin E. *The Myths of TET: The Most Misunderstood Event of the Vietnam War* (Lawrence: University Press of Kansas, 2017)

Nolan, Keith William. *Into Cambodia: Spring Campaign, Summer Offensive, 1970* (Novato, California: Presidio Press, 1990)

_____. *Into Laos: The Story of Dewey Canyon II/Lam Son 719* (Novato, California: Presidio Press, 1986)

Rapport, Leonard and Northwood, Arthur, Jr. *Rendezvous with Destiny: A History of the 101st Airborne Division* (n.p.: 101st Airborne Division Association, 1948)

Santoli, Al. *Everything We Had: An Oral History of the Vietnam War by Thirty-three American Soldiers Who Fought It* (New York, New York: Random House, 1981)

Stanton, Shelby L. *Vietnam Order of Battle* (Washington, D.C.: U.S. News Books, 1981)

The Soldier's BCT Handbook, Department of the Army Pamphlet PAM 21-13 (May 1969)

Tolson, John J. *Airmobility in Vietnam: Helicopter Warfare in Southeast Asia* (New York, New York: Arno Press, 1981)

Tucker, Spencer C., ed. *The Encyclopedia of the Vietnam War: A Political, Social, and Military History* (New York, New York: Oxford University Press, 2000)

Wiknik, Arthur, Jr. *Nam Sense* (Philadelphia, Pennsylvania" Casemate, 2005)

Zaffiri, Samuel. *Hamburger Hill: The Brutal Battle for Dong Ap Bia, May 11-20, 1969* (New York, New York: Ballantine Books, 1988)

PERIODICALS

Rendezvous with Destiny (Winter/Spring 1971)

Schuster, Carl O. "M79 Grenade Launcher," *Vietnam*, 30 (June 2017): 20.

INTERNET WEBSITES

1/12th Red Warriors Vietnam Association website (http://redwarriors.us/History1970.htm)

101st Airborne Division Vietnam Veterans Organization website (http://www.angelfire.com/rebellion/101abndivvietvets/).

"Airmobile Operations in Support of Operation Lam Son 719 - March 20, 1971," Headquarters, 101st Airborne Division (Airmobile) in The Vietnam Center and Archives, Texas Tech (http://www.virtual.vietnam.ttu.edu/cgi-bin/starfetch.exe?KkufYUz@TOqERR@MQsQIKkJ6ctA6uhrC@Y7IpmXs82xjq.H6N-qgMjUvBjShShHBnoqPkbYhRI4BM2M86fT0hHMKl2a@b9YkqQGmJrRP.RvVzIx1En35ovg/2960107001a.pdf)

Dominique, Dean. "Gun Trucks: A Vietnam Innovation Returns," *Army Logistician* (January-February 2006) [http://www.almc.army.mil/ALOG/issues/JanFeb06/gun_trucks.html (website)]

Ghost Riders (Co. A, 158th Aviation BN) website (http://ghostriders-online.org/index1.html).

Khe Sanh Veterans website (http://www.khesanh.com)

National Archives, Military Records website (http://www.archives.gov/research/military/vietnam-war/casualty-statistics.html)

Phoenix Nest (Co. C, 158th Aviation BN) website (http://www.phoenix158.org).

Rakkasan Association website (http://www.rakkasan.net/index.html) and military unit history (http://www.military.com/HomePage/UnitPageHistory/1,13506,100793%7C778079,00.html)

Rakkasan Year Book 1970 website (http://www.angelfire.com/nc/wdd101/Rakkasans2.html).

Selective Service website (http://www.sss.gov/lotter1.htm)

Smith, Yvette. "Retired Sergeant Major Soldier for life," (October

18, 2013) Official U.S. Army website (http://www.army.mil/article/113451/Retired_Sergeant_Major_Soldier_for_life/)

The New Georgia Encyclopedia (http://www.georgiaencyclopedia.org/nge/Article.jsp?id=h-590)

U. S. Army Transportation Museum website (http://www.transportation.army.mil/museum/transportation%20museum/harden.htm)

USO website (http://www.uso.org/uso-entertainment-history.aspx)

Vietnamese Embassy in the United Kingdom website (http://www.vietnamembassy.org.uk/climate.html)

PRIVATE PAPERS

Matthews, Edwin LeBron. Personal Correspondence, 1970-1971

_____. Personal Military Records, 1970-1971

_____. Personal Photographs

Matthews, Pamela C. "Returned Mail," Unpublished Account (June 2013)

ABOUT THE AUTHOR

LeBron Matthews is a retired Baptist pastor. He earned Bachelor of Science and Master of Architecture degrees from Georgia Tech, and Master of Divinity and Doctor of Theology degrees from New Orleans Baptist Theological Seminary. He has been writing in biblical and theological publications for over 30 years. LeBron is an amateur historian, especially of American military history. He was one of the founding members of the American Civil War Railroad Historical Society. His passion for historical research undergirds his first novel, Tides of War: A Novel of the American Civil War. LeBron served in the Vietnam War as a combat infantry soldier, first with the Red Warriors of the 4th Infantry Division and then the Rakkasans of the 101st Airborne Division. He is an active member of the 101st Airborne Division Association and the Rakkasan Association. He and several surviving buddies from his platoon in Vietnam maintain regular contact with each other today. He currently lives in Columbus, Georgia, with his wife Pamela. They have four adult children and seven grandchildren.

www.ingramcontent.com/pod-product-compliance
Lightning Source LLC
Chambersburg PA
CBHW030315100526
44592CB00010B/444